Of Rock and Rivers

Of Rock and Rivers

*Seeking a Sense of Place
in the American West*

Ellen Wohl

UNIVERSITY OF CALIFORNIA PRESS

Berkeley   Los Angeles   London

University of California Press, one of the most distinguished
university presses in the United States, enriches lives around
the world by advancing scholarship in the humanities, social
sciences, and natural sciences. Its activities are supported by
the UC Press Foundation and by philanthropic contributions
from individuals and institutions. For more information, visit
www.ucpress.edu.

University of California Press
Berkeley and Los Angeles, California

University of California Press, Ltd.
London, England

Library of Congress Cataloging-in-Publication Data

Wohl, Ellen E., 1962–
    Of rock and rivers : seeking a sense of place in the American
West / Ellen Wohl.
        p.   cm.
    Includes bibliographical references and index.
    ISBN 978-0-520-25703-0 (cloth : alk. paper)
    1. River channels—West (U.S.)   2. Rivers—West (U.S.)
3. Geological time.   4. Geology—West (U.S.)   5. Human
ecology—West (U.S.)   6. Landscape changes—West (U.S.)
I. Title.

GB561.W64   2009
508.78—dc22   2008040510

Manufactured in the United States of America

18   17   16   15   14   13   12   11   10   09
10   9   8   7   6   5   4   3   2   1

For my mother,
Annette Wohl,
and in memory of my father,
Richard Wohl.
Together they first brought me to the West
and taught me to look, to appreciate, and to question.

Eastward I go only by force; but westward I go free.
Henry David Thoreau, "Walking"

CONTENTS

ILLUSTRATIONS

PREFACE

Referring to Hawthorne's novels *The Scarlet Letter* and *The House of the Seven Gables,* Anthony Trollope said, "In one case he had to write it *[Scarlet Letter],* in the other he had it to write." I had to write this book. I am one of those people with an intense attachment to place—in my case, the whole interior of the American West from the hundredth meridian to the crest of the coastal ranges that form the first divide for moist air moving inland from the Pacific Ocean.

I share my attachment to the Intermountain West with numerous people. Many artists—novelists, essayists, historians, poets, painters, and sculptors—have celebrated this landscape. Immigrants come here seeking health, new economic opportunities, and a fuller life. Artists come seeking inspiration and insight. Capitalists seek natural resources they can turn to profit. Acolytes seek spiritual sources to revere and protect. Many of the critical ideas that have defined American consciousness originated in the West or in reference to it: Manifest Destiny and pioneering;

private access to public resources; national parks and forests; the spacious and unique qualities of the American landscape and American wildlife.

Various writers mean different things when they use the word *landscape*. I use the word to describe environments that are not primarily urban, although they can be agricultural. Most often, when I think of a landscape, I think of a physical and biological environment prior to or apart from extensive human alterations, even if this is an illusion because the human alterations have occurred but are not readily apparent. The dichotomy between human and "natural" inherent in this definition can foster misperceptions, but I think that many people would define landscape similarly. Perhaps this is a legacy of landscape paintings from past centuries that depict small humans amid an extensive natural world.

Within a few years of moving to Arizona from Ohio, I realized that my earlier conception of the West formed a very poor fit with my increasing knowledge of the West, and that I was not alone in my misconceptions. I grew up with traditional Western stories of self-reliant pioneers discovering a new world untouched by human influence, and with a belief that substantial remnants of that new world still existed. These stories shaped every attitude I held toward the West, from my admiration for the pioneer ideal, to my protectiveness toward natural landscapes and belief that existing national parks preserved completely natural landscapes, to my understanding of Native Americans. When I moved to the West, I gradually realized that it was neither a region unmodified by humans prior to European settlement, nor an entity that could exist apart from human management. Western history as I had learned it had never really occurred.

My sense of place derives primarily from my perceptions of the geology, climate, ecology, and human use of resources that shape a particular landscape. I knew I wanted to become a research scientist when I moved to Arizona shortly after graduating from high school. My subsequent choices of geology as my scholarly discipline, and then geomorphology as my area of specialization within geology, have strongly influenced my changing perceptions of the western United States. Geology literally broadened and deepened the manner in which I perceive landscapes. Understanding the geologic context of a mountain range implies understanding the interactions of continental-scale tectonic plates over tens to hundreds of millions of years. As a geologist, I perceive the mountain range not as a picturesque feature that just happens to be in a particular location but as part of a much broader pattern that reflects the history of the planet and that can be more fully understood within the context of this history.

If I investigated structural deformation of rock layers or geophysical processes deep within the Earth's interior, I might not pay so much attention to how climate interacts with geology to sculpt landscape or to how human use of resources alters landscape. I pay attention to climate, ecological communities, and human resource use, because geomorphology focuses on processes that shape landforms such as coastlines and river basins, and landforms reflect the interactions among climate, geology, plants and animals, and other factors.

As a river geomorphologist, I find it largely impossible to ignore human influences on landscapes. Very few regions of the Earth's surface have not been altered by human actions within the past few hundred years, even if these alterations are not apparent at first glance. And as scientific consensus has built that people are changing global atmospheric circulation patterns and climate, it is increasingly clear that no place on Earth can be considered truly free of human influence.

Time is also a critical component of understanding landscapes. Individual landforms do not respond immediately if climate or tectonic uplift changes. Each component of the landscape has inherent thresholds that govern the manner and speed with which it responds to change. Sometimes the rate of response can be very slow. Tectonic uplift can result in the formation of bedrock waterfalls that persist in a river for millions of years. Even a river formed in sand and gravel can take hundreds of years to respond to forest clearing and the spread of agriculture, as scientists have documented in regions as diverse as the Polish Carpathians and the lands of ancient Greece. As a geomorphologist, I can never look at any landform and assume that it represents a perfect balance with contemporary climate, tectonics, or land use. Instead, I must always be aware of historical changes and of how the landform might still be responding to those changes. In many of the areas where I have worked, these "historical" changes go far back into prehistory but also include patterns of land use within the past century.

My intent in these essays, and what I believe to be the unique contribution of the essays relative to the work of others who have written about the American West, is to explore how the American cultural legends of the West shaped my initial perceptions of the landscape, and how those perceptions changed during my residence in different parts of the West while I became a geomorphologist. The first essay in the collection describes the influences that initially shaped my perceptions of the West while I grew up in Ohio. The next six essays trace the gradual changes in these perceptions as I moved to Arizona and began my training as a geologist, and then moved on to Colorado to begin my professional career. The subsequent six essays explore my deepening knowledge of the West during the nearly twenty years that I have lived in Colorado and conducted geomorphic research throughout the West. The first five of these essays address specific interactions between people and natural resources in the West by examining the effects of wildfire suppression, snowmaking, livestock grazing, river restoration, and angling on streams. The final essay anchors this section by tracing the paths of water from the snowpack in the Colorado Rockies downstream to the western Great Plains and exploring some of the effects of land use on hillslopes and streams en route.

Some of my work involves basic research that seeks to uncover how interacting forces such as the hydraulic force of flowing water and the resistance of bedrock to erosion shapes river channels. A great deal of my work also involves applied research that is focused on how human use of various resources alters hillslopes and streams. The applied research in particular has gradually led me to the conviction, noted earlier, that no place on Earth remains completely unaltered by humans, and that negative human influences on landscape are accelerating alarmingly as global population continues to explode. My work as a research scientist has fostered my tendency toward environmental advocacy, because I am continually asked by resource managers and environmental groups to evaluate the effects of humans on landscape processes. I could simply state my observations and leave it at that. I have found this difficult, however, when I grow angry at what I view as unjustifiable and easily avoided patterns of resource use that have the effect of simplifying and homogenizing landscapes and thus reducing the ability of the landscape to support diverse life-forms and to recover from natural disturbances such as wildfire or floods. This internal conflict between serving as a detached scientific observer and an activist scientist is also expressed in the second group of essays. As Aldo Leopold famously stated, "The penalty of an ecological education is to live in a world of wounds."

The concluding essay integrates these diverse research experiences as I discuss my sense of place and what I think is needed to build a more sustainable society in the American West. The entire collection of essays traces changes in my sense of place that resulted from my research as a geomorphologist and from more direct experience of specific places within the West, particularly as these direct experiences contradicted stereotypes of the West.

.   .   .

This book is dedicated to my parents because everything in it represents something in me that they nourished together. I must also thank the many dear friends with whom I shared the western adventures that formed the core of this book. One great reward of friendship is the opportunity to see life through someone else's eyes, and the friends who were a part of this book have enriched my life immeasurably. Richard and Annette Wohl, Ray Kenny, Sara Rathburn and Jim Finley, Suzanne Fouty, Jim O'Connor and Karen Demsey, Lisa Ely and Keith Katzer, Dorothy Merritts, Susan Fuertsch Fanok, Deb Anthony, Doug Thompson, Dan Cenderelli and April Lafferty, and Madeleine Lecocq: thank you for being yourselves. Sara Rathburn, Suzanne Fouty, April Lafferty, and Madeleine Lecocq read earlier versions of this book, as did several anonymous reviewers. Each of them helped me to clarify the structure and function of the book, as did reviewers for the University of Chicago Press and University of California Press. Blake Edgar of the University of California Press gave the manuscript a final, careful editing, and Bonita Hurd's copyediting further clarified the writing. Thank you.

PART I

Discovering the West

The Western Reserve

I was born in the lands of the Western Reserve. In 1786 the citizens of Connecticut reserved for themselves a huge tract of land west of the Appalachians. The first European Americans to reach this tract wrote glowingly of wealth, of wetlands restless with the comings and goings of tens of thousands of waterfowl, and of rivers thick with fish. The Native Americans living at the mouths of rivers such as the Cuyahoga hunted and fished, domesticated plants, and maintained extensive trading networks. These peoples—the Adena, Hopewell, Wyandots, Hurons, Iroquois—lived well for uncounted generations amid the abundance of the well-watered eastern woodlands.

By the time I was born, in 1962, the Hopewell and their kin had vanished to the briefest mention in books and museums of regional history. The Western Reserve had long been the state of Ohio, and the Germans, Irish, Italians, and Poles thoroughly outnumbered any emigrants from Connecticut. But the names *Western Reserve* and *Iroquois* rooted the landscape in a long history of human occupation. Fifteen minutes' walk from my house, the Rocky River cut endlessly at a huge shale cliff below Old Fort Hill. Tall oaks and maples that toppled from the cliff pulled worked flint points up with their roots. If I walked the other direction, toward school, my path ran beside a cemetery where dates of 1789 or 1804 were barely discernible on the corroding limestone markers. At nearby Put-in-Bay my family visited the monument to Commodore Perry's victory over the British in the War of 1812, with its historical markers commemorating Perry's famous phrase, "We have met the enemy, and they are ours." Large grindstones chiseled out of the local sandstone for the first flour mills built along the Rocky River still lay among the undergrowth beside the river. We drove steep No Bottom Road, where daredevil drivers tested car brakes during my grandfather's youth. I perceived the landscape as the product of natural processes and the activities of generations of people.

My father gave me much of my sense of the natural history that continued to shape our landscape, as well as that of being rooted in the landscape. He was a naturalist. He taught me to identify birds and plants along the same woodland trails where he had learned their calls and blooms as a boy. We hunted for fossils in the shale cliff, and in the Cleveland Museum of Natural History we studied the sharp-mouthed skull of an ancient fish that had been found in the local shale. I tried to imagine glacial ice a mile thick above the humming green forests of summer.

My family's Sunday morning worship, no matter the weather, was a hike in the Valley. The Valley was a regional park centered on the Rocky River a short walk from our house. From the local interpretive center and museum, our path followed an old channel of the river, which by then had become a swamp thick

Summer along the trail in the Valley, Ohio

with duckweed that hid turtles in its murky depths. Beyond the swamp lay abandoned farm fields slowly growing back to woodland. Autumn winds blew in flocks of cedar waxwings, and we made frequent long pauses to watch the birds noisily feeding on the seeds of birch trees. Old Fort Hill rose steeply above the fields, its flanks partly covered in rows of pines planted during the 1930s by the Civilian Conservation Corps as a soil conservation measure. From the top of the hill we looked down on the winding course of the Rocky River.

My parents were teachers. Although as an only child I often felt put upon at being always surrounded by teachers, those dearest of teachers fostered in me a natural curiosity that knew few bounds. What began with simple rock collections and plant presses grew into roadkill carcasses bubbling on the stove so that I could boil away the bodies and reassemble the skeletons, and vials of stinky pond water lined up in the basement to be examined under the microscope. My father taught biology and chemistry, and he revered science and scientists. Our bookshelves at home included a first edition of *Silent Spring* and a copy of Darwin's *Origin of Species*. The other side of my father's respect for science was a limited tolerance for organized religion. This I imitated, and reverence for the natural world apart from humans held for me the spiritual center that a belief in God occupies for others.

As an integral part of science and scholarship came my parents' emphasis on careful observation, questioning, and testing. In northern Ohio, this inevitably led to my recognition that the natural world had for many generations been modified by humans. While I was still in elementary school, our river sampling taught me that septic tank overflows had polluted the Rocky River. My father gutted an old vacuum cleaner and converted it into an air-pollution sampler that sucked air through a square of cotton gauze. When an hour of sampling produced a soot-black or rusty-brown patch on the gauze, it was no leap of imagination to extrapolate to what our lungs must look like. I became a child activist, writing indignantly to President Nixon about pollution. A secretary sent a short, bland note on White House stationery in reply.

My indignation at perceived outrages against nature was warm, but I had no context in which to evaluate change. One spring morning I grew furious at discovering that the banks of loose shale along the Rocky River had been thoroughly reconfigured. Bulldozers! Then my father explained how a spring rain on melting snow could change a riverbank overnight.

Misunderstandings aside, my indignation at what I perceived as human trespasses against the natural landscape had grown to a deeply painful wound by the time I left Ohio. During my seventeen years of life there, the landscape that I knew intimately, and which underpinned my sense of place, was thoroughly altered. Behind our house lay a second-growth woodland and marsh. Though only a few acres, it constituted wilderness to me. Deer, fox, raccoon, woodchucks, pheasants, and songbirds inhabited the woods. From the maple saplings, I culled poles that I used to build tepees and lean-tos. I built a different shelter each summer and experimented with Native American cookery, substituting ground beef for venison. Then, when I was sixteen, our town decided to join the metropolis. The woodland and marsh were obliterated, replaced with a shopping mall, a church, an apartment complex, and a sunken freeway. I could not have been more hurt if a family member had been attacked.

The loss of the woodlands was a blow that spun me toward the West. The needle of my inner compass had pointed increasingly westward for years. Every summer my parents and I camped and hiked in Yellowstone, Grand Teton National Park, or the Colorado Rockies. While back home I had to reconstruct the Hopewell culture from a single flint tool, in Grand Teton I spent an enthralling rainy afternoon studying intricately beaded garments actually worn by Native Americans and listening to tapes of their chants. And when the rain cleared, I watched the sun set across a lake where incoming trumpeter swans formed silhouettes against abrupt mountains rising thousands of feet from the valley floor.

.   .   .

Ohio has a close-textured landscape. Rain and snow fall frequently, weathering the underlying sandstones and shales into thick, dark soils rich with the humus of fallen leaves. Trees grow luxuriantly on these thick soils: red oak, sugar maple, and beech on the uplands, sycamore, butternut, and paper birch on the bottomlands. In summer the trees form a dense green canopy loud with the trilling of insects and birds. Beyond the canopy the sky may be palest blue or gray with clouds, but its colors are always subdued and views limited by the pervasive moisture that weighs down the air. One's attention focuses on closer details. Jewelweed seeds exploding from brown jackets reveal an inner layer of turquoise blue. A stickleback minnow, sporting a sharply erect fin, noses at the layer of algae that makes the plates of shale on the river's bed slick as ice. A male cardinal perched on a holly branch forms a patch of emphatic red. By winter the colors grow even more subdued as low pressure systems sweeping south off Lake Erie dump white snow onto the blacks and browns of dried leaves and bare branches. A white-breasted nuthatch calling steadily among the dry clatterings of oak leaves may be the only sign of life while the storm lasts.

Beyond Ohio, the trees continue thickly through Indiana and Illinois. Not until Iowa does the generous spread of trees begin to contract into smaller patches, as the green of cornfields replaces the green of forests. A century ago the green was tallgrass prairie, where big bluestem sent its seed stalks eight feet in the air and its roots nine feet down.

As the forests contract, the sky grows. It remains an indistinct blue in Iowa, but the wider views allow more scope to observe the formation and movement of clouds. Thunderstorms become dramas of the sky that move swiftly to a grand climax of rain and hail and then pass as swiftly away.

At some point between Iowa and Nebraska or South Dakota, my family crossed an invisible boundary during each summer journey westward. Beyond that boundary we were in The West.

The West was a vaguely defined region where cowboys and Indians still lived. The big wild animals had not been hunted to extinction there, and the air and waters were not fouled with pollution. The West was Nature and Truth.

I visually identified the West as the place where the sky became deeply blue and the vegetation grew so sparsely that there was no limit to the view of the sky. As the air grew clear of moisture, everything looked more distinct and vivid under the strong sunlight. Many scholars believe that the aridity of western North America is one of the most important controls on the landscape and on human interactions with the land. To a new immigrant, the dryness appears first in the sense of vast spaces and vibrant colors.

Once across the boundary of the West, people identified themselves with Hollywood-style cowboys on the motel signs, or billboards inviting you to visit the local "Boot Hill." The story of the heroic West of cowboys, Indians, and mountain men was prepackaged and ubiquitous when I was growing up. I absorbed it from *Gunsmoke* and *High Chaparral* on television, and from Owen Wister's *The Virginian*.

Hollywood had much to do with shaping my imaginings of the West, but Hollywood was only the latest in a long tradition of image creators. Thomas Jefferson had interpreted the size of North American mammals as a sign of the New World's virility, J. Hector St. John de Crèvecoeur had described the American as a new man, and Horace Greeley had urged citizens to go west: every emigrant to the West was steeped in the story that America was the land of new beginnings in a pristine world of natural abundance and freedom, and that the West was the ultimate expression of America.

Or should I say myth? *Myth* comes weighted with the implication that the story is untrue. My friend Madeleine told me of her experiences during several summers as a backcountry ranger in the Rawah Mountains of Colorado. Often alone in dangerous weather and rough terrain, she realized that the heroic pose of complete self-reliance in the wilderness—the classic stuff of traditional western stories—was one of ignorance, isolation, and vulnerability. Unlike the Ute Indians who had once lived in the Rawahs, she could not live off the landscape, because she could not rely on the accumulated knowledge of culture developed in that place, or on the differing skills of others.

Madeleine's story made me wonder why we develop myths. Why did I believe so firmly in the mythic pristine wilderness of the West? Why have most other Americans? I can think of several reasons. If the land was pristine, the Native Americans were not using it fully. And if they were not using it fully, their presence didn't count and they could be forcibly displaced. If the land was pristine, American society had a chance to start fresh and create something better and purer than European culture. If the land was pristine, then dominating the threatening climate and wild animals would prove the European Americans to be strong and intelligent. Ultimately, if the land was pristine, the European Americans could be pioneers. They could tame, reclaim, and civilize the land in the Judeo-Christian tradition of mastering natural forces. If the land was not pristine . . . well, there went a large portion of our national identity.

For me, the historical existence of wilderness, and its contemporary remnants in the West, was a key to what made the West different from the rest of the United States and from Europe. Because a natural world apart from humans formed the center of my religion—the source of my inspiration, insight, and knowledge—and because western landscapes were so sparsely settled and looked so unaltered, I believed that a natural world existed in the West.

The idea that wilderness persists in the West was also an underpinning of my scientific research. Many scientists who study "natural systems" prefer to focus on examples that do not seem to be altered by humans. The aquatic ecologist seeking to understand interactions among insect and fish species would rather work in a river where species have not been extirpated by pollution or overfishing. When I measure downstream changes of river forms in relation to discharge, I do not want the complications of artificially stabilized banks, or stream flows decreased by dams or diversions. It is scientific human nature to seek out the simplest examples first, and to add layers of complexity once we understand the simple systems well enough to predict their patterns and trends through time. National park- and forestlands in the West often provide the simplest settings possible, in that fewer contemporary land uses may be occurring on these lands. In this context, evidence that beaver trapping or timber harvesting more than a century ago created persistent alterations in stream channels can be unwelcome news.

.   .   .

My parents and I moved to Arizona when I graduated from high school. I was very conscious of parallels with earlier emigrants as we drove west across the Great Plains and the Rockies. The myths of the first European American pioneers governed their actions, and our history. The pioneers displaced and killed Native Americans. Some of them grew a native arrogance based on the conceit that Americans have purer morals than other cultures. They saw no need to conserve or foster the seemingly limitless abundance of America's natural resources, and they worked through many of these resources in a few decades. As pioneers, they took no lessons from predecessors, and made mistake after costly mistake as they let the precious topsoil of the shortgrass prairie blow away, or stripped the hills of trees and of the ability to store and gradually release water. We later inhabitants of the American landscape continue to put credence in those myths, and to twist them for our own ends. Western historian Bernard De Voto noted that the great danger of the Western myth of the lone pioneer or cowboy is that it can be manipulated to romanticize exploitive, destructive business practices, and the mining of timber, soil, and water. As De Voto wrote, "The trouble is that 1880 [perceived endless resources] is dangerous to the West, for it damages land. And to damage land is to destroy water."

What I did not realize was how closely I paralleled the nineteenth-century pioneers in my stock of preconceptions. I journeyed West with a naturalist's appreciation of how geology and biology shape a landscape, and an environmentalist's concern about how humans alter the natural order. I relied on the empirical knowledge of science and had a religious reverence for the natural world as something apart from and above humans. I believed that nature had been unspoiled, because unaltered by humans, prior to the arrival of European Americans, and that it closely approximated this state in the American West. And because nature was my spiritual center, I believed that to move West was to give myself a better life in a new world. Like Thoreau, by going westward I would be free.

Nineteenth-century pioneers littered the way west with discarded baggage. The Great Platte River Road and the Hastings Cutoff across the Humboldt Sink were marked by furniture left behind overloaded wagons. The perceptive continued to discard their baggage once they settled into the new landscape, replacing prejudices against Indian savages with respect for Native American knowledge of place. Both the physical and the metaphorical discarding came hard.

My baggage is equally persistent. More than two decades of living in the West has changed my perceptions of the western landscape considerably. It has started me on the path toward recognizing that landscapes and ecosystems are not so much discrete entities as ongoing processes of change that have been everywhere affected by humans for thousands of years. The history, as opposed to the romantic myth, of the West has been difficult to absorb. The West is not a pristine wilderness, for people have been altering natural processes here for millennia. This apparently simple fact has for me had an enormous repercussion: I recognize now that there is no unspoiled source of truth or life. There is no western reserve. And this makes all the difference.

So where does this leave me and others who have assumed that wilderness must be protected in isolation? The isolationist view of wilderness may have contributed to American society's contemporary isolation from the land. Yet skyrocketing human population and material consumption are overwhelming the land's ability to sustain us, suggesting that we need to buffer some landscapes from current human practices as much as possible in order to simply sustain ourselves. We in the contemporary West live increasingly segregated lives, with cities at the front and back doors and "wilderness" a drive away in the national parks and forests. I would be integrated. But how? I am searching for that, making up a new story as I live in the West and as I write these essays.

The rest of these essays explore in detail this process of retelling the story of the West. The process itself is intensely personal and yet also national, for the story of the frontier as exemplified in the American West has been repeatedly identified as a source of our national identity, by thinkers from Thomas Jefferson to Frederick Jackson Turner and Wallace Stegner. As people all over the world struggle toward means of sustaining human life on this planet beyond the twenty-first century, it is imperative that we understand where we have come from and the stories that have guided us in the past. Only by carefully examining the assumptions implicit in past stories can we tell new, and perhaps better, stories.

A Sense of Space

Geology here forever dominates life and gives it its ultimate meaning.
Frank Waters, *The Colorado*

My parents and I moved to Arizona in June 1980, and I had one summer free of the restrictions of schoolwork before I began my studies at Arizona State University. The desert landscape and its inhabitants had been curiosities during short vacations. Now they became the context in which I lived, and I explored them eagerly. I had lost my Ohio favorites: the chickadees, cardinals, and goldfinches. But I gained whip-poor-wills calling *whirl-pool, whirl-pool* in soft, deep tones, and little gray-and-white mockingbirds that sang and clucked and screamed in an unpredictable succession of notes. Early in the morning, coming out onto the shadowed north patio, the cement cool against my bare feet, I watched the Gambel quail chase each other across the yard. Their bodies upright as though unmoving and their thin legs blurred in motion, the quail resembled mechanical toys. Each male had a curved shaft of feathers like the upper half of a question mark on his forehead. Their querulous two-part call, repeated over and over, soon came to evoke the desert for me.

I absorbed the details of my immediate surroundings first, observing the plants and animals in my neighborhood. Big, angular jackrabbits and smaller desert cottontails scattered gravel with their powerful hind legs as they crossed the xeriscaped yard. Sometimes I'd see a roadrunner vigorously dispatching one of the little gray lizards that lived beneath the agave and around the date palm. More often, the lizards were left in peace to do their endless push-ups on the back stone wall.

I learned the desert ways quickly, exercising during the brief cool of early morning. As the day progressed and the temperature climbed to triple digits, the neighborhood grew as silent and unmoving as a ghost town. Like everyone else, I napped until the daily drama of a thunderstorm ruptured the heat.

Each afternoon from mid-July through August, storm clouds swept upward into immense piles over the mountains surrounding the valley. Thunderheads, gold and gleaming white, massed together like thick-petaled, full-blown roses arranged against the flat, pale blue sky. Then the tufted heads of the slender palm trees bent almost double as the storm rushed down into the desert basin. The sky grew dark purple and the air smelled musty with the approaching storm. The dust front came first, rolling and boiling, blotting out the sun to create an eerie half-dusk. Dust seeped in through the most tightly closed doors and windows, leaving a gritty feel in my eyes and mouth. I could smell the rain coming behind, and a broad blue sheet of virga hung in a curtain from the sky, streaked by violent, hurried lightning. The rain came in fat, sploshy drops pelting down so fast that streets and then dry stream channels flooded. Within a day or two of the first rain, the desert sprouted a quick fuzz of short green grasses and forbs.

Growing up in the tornado belt, I'd been familiar with the energy of storms. But the tornadoes, thunderstorms, and blizzards of northern Ohio did not have the same implications for landscape processes as the summer rains of Arizona, because moisture was seldom a limiting commodity in Ohio. In Arizona, landforms and living creatures waited out the bust of dry periods, only to undergo rapid change with the boom of the rains. Flash floods coursing down the normally dry streambeds could move boulders that might wait a decade or more to be transported by the next high flow. Sediment that had slowly accumulated on hillslopes through the weathering of bedrock could be swept downslope in a debris flow lubricated by rainwater. Some portion of the water moving furiously across the ground surface would infiltrate deep into the sediments at the base of each mountain range, contributing to the aquifer that provided critical water supplies to municipalities and agriculture. What did not infiltrate or evaporate contributed to a sudden growth spurt among the desert plants.

Studying desert botany, I quickly realized that most plants based their survival on the absent or the cryptic. In Ohio the plants competed for sunlight, squeezing together in every patch of ground. In Arizona the plants had plenty of elbow room and sunlight, but they needed to somehow save every available drop of moisture. Stay dormant till rain and then grow fast: that is one adaptation to the desert's heat and aridity. Annuals such as lupine live their entire life cycle in a week, then leave seeds that lie dormant in the soil until rain brings enough water to wash over the seeds and dissolve the chemical coating that inhibits their growth.

But other botanical innovations abound: Mesquite trees send taproots tunneling as much as a hundred feet down to subterranean water. Saguaro spread a broad web of shallow roots thirty to fifty feet beyond the plant to soak up water whenever rain falls, then store the water in expandable spongy tissues. As in other cacti, the saguaro's leaves are replaced by spines, and the plant photosynthesizes through its green stem. Jojoba grow a waxy coating to keep the desiccating winds from sucking moisture out of their leaves. Sagebrush use a layer of tiny, furlike hairs for the same purpose. Palo verde reduce their leaves to tiny green nubbins, which they lose altogether during the dryest times, relying on their green branches and trunk for photosynthesis. The tiny leaves of ocotillo also fall from the plant during the drought, but the ocotillo becomes dormant, living off water stored inside the plant. Creosote and brittlebush extrude toxins from their roots to keep other plants from growing too close to them, a strategy that reminded me of growing Western cities buying up the water rights from all the neighboring farms.

Plants dominated my perception of landscape in Ohio. They were immediately present and apparent to my childhood eye, and they constituted the scenery. Yet in some ways I never really saw the plants in Ohio. I identified and appreciated specific plants, but I never thought too much about what made each species unique. The desert plants of Arizona were so different that almost immediately I made the connection between the Earth spinning through space, tipping toward and away from the sun, creating the seasons and changeable weather that provide the challenges to which plants must adapt. Ohio plants had been for me just plants. I had taken them for granted. When I realized how Arizona plants contorted themselves to obtain water, I understood that

Saguaro in Arizona

The spines of a saguaro

plants everywhere evolve in response to competition for limited resources, whether the limitation comes from water, sunlight, nutrients, or temperature. The desert helped me to see both the forest and the trees.

The desert also helped me to see the rocky skeleton of the Earth. Dryland plants form such a thin veneer over the topography that, in my freshman year at Arizona State, my enthusiasm to comprehend the landscape quickly drew me beyond botany to the underlying geology.

Geology is obvious in the desert. You have to close your eyes, literally and metaphorically, to miss the rugged mountains surrounding the Phoenix basin. To the east the myth-haunted massif of the Superstitions rears abruptly in cliffs of pale orange tuffs towering over dark basalt flows. Hopeful miners still grow confused searching for the wealth of gold rumored to be in the Lost Dutchman Mine among the ridges from which Weaver's Needle rises like an exclamation point. To the west stands the thin spine of the White Tank Mountains, where rains spread coarse chunks and crumbs of weathered granite in broad aprons across the mountain's foot. The dark metamorphic rocks of the Sierra Estrellas form the southern boundary of the Phoenix basin, and the orange, pockmarked rocks of Camelback Mountain and Papago Buttes rise abruptly in the middle of the city. To the north the Bradshaws and the Vultures climb gradually up toward the Colorado Plateau.

Having chosen geology for my profession during my freshman year of college, I still had to learn how to perceive the implications of geologic history for any particular landscape. I do not think anyone can really comprehend geologic time. Metaphors abound. Extend your arms straight out on either side to represent geologic time. Trim your fingernails, and you remove human history. Or think of a clock ticking through a twenty-four-hour cycle. At just before a minute to midnight, hominids appear, and at a few seconds before midnight they evolve to *Homo sapiens*. Such metaphors provide perspective, but who can really conceive of 4.7 billion years? Not I. The closest I can come is to understand the alternating fast and slow pacing of changes in the Earth, and to realize which features are ancient and which are fairly young.

My perceptions of deep geologic time and change began with the geology of southern Arizona, which records a tumultuous history. As the Pacific tectonic plate is forced down into the Earth's mantle beneath the western edge of the North American plate, the edges of the plates are compressed as though in a giant vice. The rocks behave plastically as they are folded into mountains under great heat or pressure, or they rupture in a brittle fashion along faults. Heat and pressure melt the rocks as the Pacific plate is forced downward. Some of this melted material bubbles up through the overlying crust as volcanoes like the Superstitions. Some of it rises as superheated water carrying dissolved metals. The water cools as it is forced into cracks in the overlying rock, and the metals collect along the cracks to form veins of gold and silver ore. It was these veins that attracted many of the first European Americans to Arizona.

I learned the regional geology in field trips spread over four years. Field trips were one of the great strengths of the undergraduate geology program at Arizona State University. Every course I took involved at least one-day field trips, and many of them had multiday trips. These were the times when I learned to use the rock hammer bequeathed me by my father, as well as a compass and air photos. I learned to identify rock units and to map their structure. I shredded various parts of my clothing and my anatomy on tough rocks and the tough plants trying to cover them.

One semester we visited the Superstition Mountains and learned of the violent history that had created the abrupt landscape. A year later we visited the Estrellas and learned how to identify the suite of metamorphic minerals that recorded the conditions at the time of crystallization like so many little temperature and pressure gauges. The desert appears static during a short visit. I seldom saw a tree knocked down or a riverbank collapsing from erosion, as I might have during a hike in Ohio. But integrating the knowledge from the various field trips taught me that the apparently changeless landscape was actually the product of immense and continuing changes.

The Pacific plate has been moving downward beneath the western edge of North America for more than six hundred million years. The rate and style of downward movement have varied during this huge span of time, but the Pacific plate has never made a right angle bend downward at the junction. Instead, the plate moves downward at some lesser angle, so that magma rising from the melting plate may surface far inland from the west coast of North America. The location of the west coast has also moved westward with time, as bits of colliding plates have been accreted to the edge of the continent. The changing configuration of the plate boundary has caused corresponding changes in the geology of Arizona.

Many of the southern Arizona ranges have a core of rocks 1.5 billion years old that formed when molten material rose from the mantle to form mountains. These mountains were gradually worn down, their weathered rocks spread as sediments across basins between the ranges. Then the landscape was rejuvenated as new magma was forced upward, folding and faulting the existing rocks and raising new mountain ranges, again and again.

The compression and rising magma were replaced by tension. Southern Arizona lay between forces moving part of the North American plate west while the rest of the plate moved east. A line of mountains rose like a diagonal slash west to southeast across the map of Arizona. A ring of volcanoes erupted along the western Phoenix basin. Huge explosions left collapse pits and belched thick clouds of red-hot gas that rolled outward to deposit ash and fractured rock. Another pulse of magma domed up the central part of the collapse pit to form today's Superstition Mountains. Slow outpourings of basalt carpeted the region with layers of black lava. A new violent eruption produced more clouds of hot gas and ash that cooled to tuffs.

Within the last two million years, the blink of a geologic eye, Arizona experienced a new phase of volcanism as basalt lavas flowed across large areas. The basalt flows produced the dark brown plateaus so characteristic of the high desert in central Arizona. This volcanism coincided with the continental-scale glaciers advancing and retreating to the north. Glaciers in Arizona were confined to the highest elevations, but even the desert had periods of cooler, wetter climate during glacial advances. When the climate grew cooler and wetter, plant species from higher elevations moved downward toward the basins. When the climate warmed and dried, the desert plants took back the terrain they had lost.

Humans reached Arizona at least twelve thousand years ago, as the climate warmed once more following the last ice age. Their descendants formed the tribes Anglos know as Hopi or Pima, who were partially displaced by aggressive Navajo and Apache entering the region circa A.D. 1100. The most recent displacement came with the European Americans circa A.D. 1700. Spanish conquistadores ranged across the desert in a vain search for golden cities, and Spanish friars began small agricultural settlements and attempted to convert the Native Americans to Catholicism. And, as the fever of Manifest Destiny gripped Anglo-Americans, they sought beaver furs and gold in the Arizona mountains, then stayed to develop crops and ranches along lines defined by the precious water of rivers.

The history of Native Americans and Spaniards is much more recent and tangible in Arizona than is the analogous history of Native Americans, English, and French in Ohio. Arizona is rich in national monuments, and here I could easily and literally touch the past: adobe walls melting back into the desert at Casa Grande; carefully placed rock walls forming vividly hued Cubist architecture at Wupatki or Tuzigoot; whole villages fitted into cliff-face caverns in beautiful continuity. At Tumacacori and San Xavier del Bac, I admired claret-colored strings of chile peppers drying against tan adobe walls, and took shelter from the desert's heat on shadowed walkways framing interior courtyards green around softly flowing water. Willa Cather's novel *Death Comes for the Archbishop* had predisposed me to admire the Spanish friars. The enduring beauty of their architecture strengthened my admiration. And I read accounts by Martha Summerhayes, Herbert Young, and Sharlot Hall of life in Arizona during the last decades of the nineteenth century and the start of the twentieth century. My grandmother had already begun her life during the time period covered by these descriptions of life on the Arizona frontier.

Learning the desert's geologic and human history was like seeing a time-lapse movie of landscape change. I imagined mountains rising and being worn down, seas advancing and receding. At a closer focus, I imagined the processes by which the mountains rose and fell: the volcanism and the faulting that sent shock waves of earthquakes in all directions, interspersed with the rockfalls and landslides that brought the mountains down. I imagined Paleolithic hunters pursuing mastodon across lush savannas, and cotton farmers diverting the flow of rivers across their fields. If I could not truly comprehend deep geologic time, I could at least appreciate that geologic processes were ongoing, and that humans interacted with these processes in complex ways. The contemporary landscape became but one frame of a constantly changing movie. Probably the most important lesson I learned during my first years in the desert was that few shared my appreciation.

.   .   .

During an environmental geology field trip, we visited the margins of the Phoenix basin. Here at the transition zone, irregular knobs and ridges of bedrock extend out from the mountains into the basin. The bedrock protrusions are largely buried beneath a thick mantle of sediment. Water, temperature changes, and the organic acids excreted by creatures ranging from microbes to trees inexorably weather any bedrock exposed in the mountains, decomposing the solid rock into fragments carried down to the basin by wind and water. Every good rain brings down a new load of sediment in a flash flood or a debris flow, building broad alluvial fans across the mountain front and burying the bedrock protrusions.

Some of the water soaks into the ground and remains in the interstices between the coarse sand and gravel, forming a rich underground reserve in this land of little rain. Each year's addition of rainwater to this underground aquifer may be a tiny increment, but the water can remain in place for thousands of years. The groundwater once lay close to the surface in places and fed the larger streams that ran year-round.

When the first European Americans settled in Phoenix, they built a flour mill along the banks of the Salt River where beaver dams ponded water in which malarial mosquitoes bred. The settlers diverted the stream water into irrigation canals, as had the Hohokam people more than five hundred years earlier. As the irrigation water evaporated, it left behind salts that gradually rendered the soil infertile, forcing the Hohokam to abandon their extensive network of canals. European Americans diverted so much water to their thirsty crops that the basin's water table began to drop. The rivers dried up, becoming intermittent pools during much of the year, then ephemeral channels that flowed only after rain.

Once the supply of surface water was gone, the settlers sank wells to pump the groundwater. They sucked water from the underground aquifer faster than it was replaced by rainwater. As each drop of water was pulled out, the surrounding sediment grains collapsed into the resulting cavity, and the basin surface subsided. The subsidence occurred unevenly, particularly along the edges of the basin over the buried bedrock ridges. Tension cracks appeared at the ground surface, first as little cracks you could cover with your hand, then as gashes in the earth large enough to swallow a car. Some of the local residents filled them with abandoned cars and other trash.

The basin margins are also prime building sites, because they lie above the city smog and offer long views. On some geology field trips, we toured new housing developments where developers had bulldozed earth into the subsidence cracks, then built and sold new houses on the site. When cracks appeared in the foundations and walls of these houses a year or two later, the developers denied any responsibility. The homeowners did not think about cracks until they appeared, being unaware of the rapid changes occurring in the basin sediments because of groundwater withdrawal.

Some of the very expensive houses also backed onto picturesque collections of huge boulders and even incorporated the boulders into their landscaping. What seemed to be missing was recognition that the boulders had been produced by rockfall, an episodic but ongoing process. Similarly, houses perched next to shallow dry channels suggested that neither builder nor occupant realized that those channels would fill, overflow, and perhaps erode their banks and move sideways during any rain heavy enough to generate runoff.

My dawning awareness of landscape change did not yet include a real understanding of human-induced change, despite the examples I had witnessed. It was easy at age eighteen or nineteen to dismiss the subsidence cracks, drying stream channels, and rockfall or flash-flood hazards as isolated examples. My environmentalist ire was provoked by learning that businesses such as insurance corporations flood-irrigated fields of notoriously thirsty cotton at an economic loss, wasting precious water and using the endeavor as a tax write-off. But this too remained in my mind as a single example of shortsighted greed. I did not yet see the larger context of thorough landscape change.

For the most part, I was very busy living my own version of the western myth. I read much regional history and literature, absorbing the stories about Spanish conquistadores and Anglo gold seekers. I perceived the Native Americans and Europeans of Arizona history as picturesque inhabitants of a vast, challenging landscape on which they had little impact, and I celebrated my inheritance of their tradition of courage and self-reliance. My parents and I enthusiastically hung paintings of cowboys and fur trappers on the walls of our new Arizona home and indulged in the regional fashion of mesquite-wood barbecues and chuckboxstyle suppers that mimicked what real cowboys ate historically. We bought straw cowboy hats for our forays into the surrounding mountains.

. . .

The mountains form punctuation marks in the desert. They rise above the broad sweep of the more homogeneous desert basins, each range distinctive in appearance. Because the basins are now largely developed for agriculture or cities, hiking inevitably focuses on the mountains. Each hike up into the mountains played with my sense of space. The trails generally begin on the broad alluvial fans at the base of the mountains, where the world seems nearly infinite. Brushstrokes of dry, sandy washes trace gracefully undulating pale bands across the darker fan surface. Green palo verde as wispy as smoke and thorn-armored mesquite line the washes. During dry periods the tree branches are bare. The desert basins have no green, easy shade, only small, hard-won patches beneath a scantily leafed bush or tree. The cottonwoods are more generous in their shade, but they need water and are restricted to the banks of streams that flow regularly.

With the rains of late winter and early spring, the desert trees quickly sprout leaves and grow masses of blossoms that delight the bees: sunshine yellow for the palo verde and mesquite, lavender for the ironwood. Twirling insects move through the soft air, and even the drier uplands between the stream channels are covered with color as the low-growing annuals and perennials spread a quick carpet of flowers across the sand and gravel. Above the layer of forbs, cacti break out in extravagantly large, vividly colored blooms. Creosote dissolve into puffy white smoke, and brittlebush covered in yellow flowers form sunbursts across the

Cacti and palo verde tree in the White Tank Mountains, Arizona

desert. The exuberance of the spring flowering in the desert is a visual thanksgiving for the gift of rain.

This gift comes in boom-and-bust cycles. When the rains arrive again in late summer, their violence sends water running in shoots and streams from every surface and tumultuous reddish brown waves quickly form in the dry stream channels. These waves sweep everything before them, sticking debris in the overhanging branches of the mesquite trees, uprooting some trees, and providing new germination sites for seedlings.

The mountains are directly responsible for the life-giving water of the desert. Their presence enhances the possibility of rainfall, because warm, moist air masses forced to rise to higher levels of the atmosphere as they cross a mountain mass cool down, allowing some of the moisture in these air masses to condense and fall as precipitation. Weathered mountain bedrock also forms the sediment that rivers and debris flows bring downhill, providing the matrix that stores groundwater.

Most trails into the mountains follow the stream channels. Hiking these trails, I lose the feeling of limitless space as the channels narrow and deepen into canyons. The canyon bottoms form ladders of rock, stepping up from pools, over ledges of small waterfalls. Summits that initially seem close retreat as the canyon twists and forks. Details absorb my attention. I linger beside quiet cups of shadow, or slowly flowing water edged with green mosses and ferns. Quail scurry in wavering lines between patches of scrub. Thrashers give their jubilant morning calls, while far above I hear the high, harsh scream of a hawk. Crack-voiced ravens glide the desert thermals over the ridges. Cactus wrens sing melodiously from spiny perches.

As the trail ascends, the vegetation changes rapidly to low woodlands dominated by juniper and pinyon pine, with fragrant clumps of sagebrush in the understory. If the mountains are high enough, the woodlands in turn give way to forests of tall pine, spruce, and fir. Streams of trees spill over the canyon rims, trickling in thin lines down the side ravines and pooling beside the valley bottom. Breezes flow rhythmically as water among the leaves of cottonwoods and sycamores clustered along the streams. From the ridges high above comes an answering breeze, slipping among the rocks and pines. This is the wind of the West, heard before felt, blowing through a pine up the slope, and then around me, softly, gently made into music by the pine needles. There is such a sense of open space to be had in hearing the wind come and then continue beyond me. Nothing to impede its progress, only the pines to tease it gently along its way.

Eventually the canyons merge into hillslopes that climb to ridge crests far above the desert. Here space once again expands to infinity. The ridges are fortresses of coolness scattered across the desert basins and, sometimes in the early morning, islands in an ephemeral cloud sea of opalescent white. There I have found bee pastures in meadows starred with yellow daisies, white asters, red paintbrush, and blue lupine. In winter I have followed northern slopes where the snow creates fleeting designs as it settles at the base of agave leaves or tips the pine boughs. I've found regions high and bright, wind-blown and mountain-rimmed, where the snow sits lightly. I've explored a place called Heart of Rocks, where the weathered volcanic spires cluster like grizzled old men and the clouds rushing past give the illusion that the rocks are moving.

There were days when dusk fell in a golden wash that set the pinnacled rock walls glowing with a light reminiscent of a Thomas Moran painting. Coyotes greeted the sunset with their weird yipping and howling until night hushed everything beneath a crystalline sky. I had nights when the dark silhouettes of pines gave a depth to the sky that is lacking in the desert, where the night sky seems close enough to touch. Other nights fell with fire raining from the sky. In the storms of August, mountain lightning has many faces. It comes as sudden blinding flashes from one point illuminating the entire sky, and it zaps down from cloud to ground in forks, in combs, in single strands. It comes as a glowing white sphere racing from cloud to cloud in erratic, elliptical paths, leaving glowing trails. Then the wind picks up and the air smells of rain.

Desert mountain summits are described as islands in the sky because their plant and animal communities are so distinct from those of the surrounding lowlands. During past episodes of cooler, wetter climate, the plants and animals of the summits were able to survive at lower elevations, allowing genetic exchange between populations now isolated in refugia. I had learned in Ohio that plant and animal communities shifted southward as the great Pleistocene ice sheets advanced, and then returned northward as the ice sheets melted. The Intermountain West was too dry to have a continental-scale ice sheet, but individual alpine glaciers advanced down mountain valleys, driving plant and animal communities before them. My geology classes taught me how to interpret the history of mountain building and erosion recorded in the bedrock, and the history of climatic and ecological changes recorded in deposits of glacial sediment and fossils of plants and animals that lived during the past few thousand years.

.   .   .

Above the geology, the sky dominates the desert landscape. So many afternoons I walked outside, my head filled with work, to be arrested by the beauty of the sky. At sunset in August the cloud bottoms swirl and weave into each other, sinuous blue and purple stains, until the rain blots them out in a hazy gray smudge. And at sunset in February, dark masses of clouds lie just above the horizon, loosely interwoven strands of rose, amethyst, and purple at their base. The clouds are thinner above, where only a few strands of steel blue remain, surrounded by the blue silk night and adamantine crescent moon. These rain clouds are plump and solid as Baroque angels, and they give depth and scale to the unending vastness of the desert sky.

Geology gave me a sense of infinite time and of the slow but ongoing change of the Earth's surface. It also gave me a sense of accelerating change as the landscape responds to human activities, and of the human ability to ignore or forget these changes. The desert landscape gave me a sense of infinite space. The first rush of falling in love with that landscape kept me from inquiring too deeply into the extent and intensity of human-induced change.

During a decade of living in Arizona, I slowly learned how the Native Americans and Europeans had lived in the vast and challenging desert: they altered the landscape to the extent that they were able. In A.D. 1600 the southwestern deserts had an estimated Native American population of 120,000 to 150,000. Hunters such as the Apache used fire extensively for game drives. Gatherers shaped vegetation communities by their harvest practices. Pueblo tribes in the Gila River watershed used low stone and earthen walls to retard runoff from rainwater. They built canals, terraced gardens, and check dams to stabilize stream channels, creating what James Hastings and Ray Turner called "a continuous ribbon of disturbance," as the Pueblo people followed laterally shifting stream channels. Pima tribes diked and ditched alluvial fans at the base of ephemeral mountain stream channels in order to plant crops. Master irrigators like the Hohokam built hundreds of miles of irrigation ditches that forked off from large rivers.

The earnest Spanish friars were part of an invading force that decimated Native American populations with disease and warfare, and introduced domestic animals that further altered desert plant communities and landscape processes. The Mexicans and Anglos who succeeded the Spanish steadily increased the density of human and livestock populations and the extent of irrigated agriculture, as well as mined and trapped fur-bearing animals.

It took many years before my appreciation of the picturesque and heroic aspects of humans occupying Arizona matured to include an understanding of the role of those humans in creating the contemporary landscape. My horizons expanded through time and space during my first years in Arizona as I gradually changed my perceptions of the primary traits that came to define the West for me. The first and most obvious characteristic of the western United States is the spaciousness produced by dry air and long views, limited vegetation, and low population density. This sense of space is often accompanied by a perception that the landscape has undergone only very limited alteration by humans because the land is sparsely populated and urbanized. Contradicting this perception of limited human effects, however, is the past and continuing use of natural resources in a manner that is economically unsustainable because the resource (timber, water, soil) is depleted faster than it can be replaced. And it is ecologically unsustainable because biological diversity and the ecosystem services such as clean water, clean air, and soil fertility on which people depend are progressively lost. I began to understand the implications of these traits that unify the West while I lived in Arizona, but I still had much to learn.

River Days

Paradise Found

During my sophomore year in college I took a course on the geology of the Grand Canyon. The main attraction of the course was a five-day raft trip down the upper half of the canyon immediately after the spring semester ended. The course was taught by Troy Péwé, a veteran teacher and river guide. I already knew the canyon fairly well, having hiked and skied along both rims. But the idea of seeing it by floating through its heart intrigued me, and influenced my subsequent decision to pursue river geomorphology.

Like many undergraduates excited by their major, I wanted to specialize, successively, in each of the geology courses I took. During my introductory geology class, I was fascinated by the idea of reconstructing ancient geographies from the grain size, mineral composition, and layering of sedimentary rocks, and I decided to focus on sedimentology and stratigraphy. Then I took structural geology and grew enthused about tectonic plates crushing together and deforming rocks into complex folds. Next it was the chemical intricacies of petrology and the ability to infer events

The Grand Canyon from the South Rim, Arizona

deep within the Earth that excited me. But in my junior year, geomorphology really swept me off my feet. Organized around two weeklong field trips, one to Southern California's coast and the Mojave Desert, and the other to southern Utah's canyonlands, my geomorphology class integrated all of my interests. To understand landforms and the processes that shape them, you must know geology, chemistry, ecology, climate, and human history. Rivers are clearly the prime shapers of many of Earth's landscapes. And who can resist the allure of flowing water? So after being primed by five days rafting on the mighty Colorado River, I chose to study the physical forms and processes of rivers as my specialty.

Our raft trip began at Lees Ferry. With four semesters of geology courses behind me, I proved to myself that I still had a lot to learn, when I missed the Chinle Formation. Péwé sent me over to the base of a cliff about half a mile away while we waited for the boats to be loaded with our gear. He wanted me to look at the Chinle, which includes a layer of volcanic ash created by an eruption two hundred million years ago. The trees knocked down and buried by that blast now form the Petrified Forest, and the ash has weathered to form the rounded hills of the Painted Desert. I went looking for a rock unit and dismissed the crumbling clay that covered the base of the cliff. The clay was the Chinle, as Péwé laughingly explained when I returned. Chastened, I climbed onto a raft. Not chastened enough, however, to keep me from nabbing the front seat in the first boat.

The deep water was green and icy cold, clear enough to see the shadows of large trout on the rippled white sand below. Heat waves shimmered between the cold water and the steep red rock walls, and each drenching in a rapids was a rude shock.

Lees Ferry is a broad, fairly accessible point along the canyons shutting in the Colorado River. That was why John D. Lee established a ferry there in 1871. The ferry crossing was a good point at which to help his fellow Mormon emigrants, and seemingly sufficiently remote to help him evade anyone who might be inquiring into his participation in the Mountain Meadows Massacre several years earlier.

Downstream from the ferry, the walls of Marble Canyon close in swiftly. Rapids punctuate the smooth downstream flow of the river wherever a tributary canyon dumps its load of boulders. I spoke the names of the rapids as we floated them—Badger Creek, Soap Creek, House Rock, North Canyon—creating a mental list of mileposts of the journey. At each bend I strained forward and snapped enough photographs to boost Kodak stock. At the bigger rapids I held on with both hands and imagined a bucking bronco as the neoprene raft doubled into a U shape beneath us.

My watch face had a crack in it, and the watch died in the first rapids. I needed that. I have always been obsessed by time and am seldom without a watch. Unless I wanted to be a nuisance by constantly questioning others, I now had to rely on sun time and river time. In river time, the cooler mornings are for exploration and exertion; the baking-hot afternoons are for naps, sipping beer, and talking quietly in the shade; and the lovely star-strewn evenings are for contemplation of the cosmos.

We began at Lees Ferry in rocks two hundred million years old composing units called the Shinarump Conglomerate and Moenkopi Formation. As we followed the river downstream, we went literally and metaphorically deeper into the Earth. The red-and-orange rock walls bordering the river grew progressively higher as the river cut downward, until they formed a canyon within the canyon. The Shinarump and Moenkopi became the upper canyon walls, and the older, underlying Kaibab and Toroweap limestones appeared along the river.

From Lees Ferry downstream to Phantom Ranch, progressively older rocks outcrop along the river. The whole vertical sequence forms a famous stratigraphic column that we had to memorize for the class. Twenty years later, my silly mnemonic "King Tut Could Have Supper Ready To Beat Most Bachelor Tutors" still comes readily to mind, and I can reel off "Kaibab, Toroweap, Coconino, Hermit, Supai, Redwall, Temple Butte, Muav, Bright Angel, and Tapeats" with a facility that impresses my own students.

Each new layer of rock changed the character of the canyon scenery. The walls grew steep and narrow where the hard gray limestones replaced the soft, blood-red Moenkopi. Steep-angled beds of ancient sand dunes etched the cliffs of white Coconino sandstone. The light-gray Redwall limestone was stained red by sediment washing down from the softer sandstones and shales of the overlying Supai Formation, but at the high-water line the Redwall turned abruptly pale. Sand suspended in the river had sculpted and polished the hard limestone into sharp-edged flutes and potholes that reflected the sunlight. Seeing this, Major John Wesley Powell named this stretch of the river Marble Canyon during his pioneering 1869 river trip.

The legacy of Powell was continually with us. Péwé quoted frequently from the major's eminently quotable journal. More than a hundred years after Powell wrote it, the journal remains standard reading for river travelers in the Grand Canyon. The man who floated the Colorado River singing the hymn "Flow Gently, Sweet Afton" from a wooden chair strapped high on the deck of a wooden boat comes alive in his narrative. Geologist Powell vividly described the ancient meeting of snowmelt and molten rock when he imagined the basalt flow that formed the canyon's mighty Lava Falls Rapids. Frederick Dellenbaugh has quoted Powell the adventurer as exclaiming, "By God, boys, we're gone!" and "Bail for your lives!" at exciting points. Three members of the first expedition mutinied and hiked out of the canyon to their deaths. Despite hardship and danger, Powell used the geologic opportunities of the canyon: "All about me are interesting geologic records. The book is open and I can read as I run." I imitated him as best I could.

To help us read the book, Péwé lectured on Powell's recognition of the concept of base level and the role of the Colorado River in making the canyon grand. The deceptively simple concept behind base level is that the downstream-most elevation of a river can influence the river's character all the way to its headwaters. If advancing continental ice sheets store water in glaciers and lower global sea level, rivers draining to the oceans will cut downward in response to lowered base level. Or, if base level stays constant but the upper drainage basin is uplifted by mountain building, the upper portion of the river will cut down to maintain its relation to base level. At a time when geologists weren't sure whether a river could cut its own canyon, Powell proposed that the Colorado River had created the Grand Canyon in only a few million years in response to uplift of the Colorado Plateau, across which the river flowed.

The rocks exposed along the river record more than one episode of uplift. One of the most readily identifiable occurs at the Great Unconformity, where steeply dipping rocks of Precambrian age are overlain by horizontal layers of Cambrian-age Tapeats sandstone. A gap of millions of years separates the Precambrian and Cambrian rocks. During this time the older rocks were folded and uplifted in an episode of mountain building, then eroded back down to gently rolling lowlands before the younger rocks were deposited above them. The Great Unconformity is impressive because you can put your finger on the abrupt boundary representing an unimaginable length of time between the tilted and the horizontal rocks. Or, in our cases, your lips. One of our rites of passage was to kiss the Great Unconformity.

Powell recognized the upheaval recorded in the Great Unconformity, and the more recent uplift, which the Colorado River was cutting into. Despite the dry climate and sparse vegetation of the Colorado Plateau, the rocks break down into sediment. The river uses this sediment as a tool to grind the rocks that it flows past.

The Spaniards named the Rio Colorado for its turbid, reddish brown waters. At Lees Ferry the river water is clear because it has just been released from the giant settling pond of Lake Powell above Glen Canyon Dam. The clear waters grow gradually more silty as each tributary adds its load of sediment, but the water does not regain its historical appearance for tens of miles downstream unless one of the major tributaries, such as the Paria or Little Colorado, is in flood and carrying a load of silt.

The canyon bottom juxtaposes fire and ice. Only two or three feet above the water, the searing summer desert air baked my nose and throat as I breathed. The canyon walls immediately above the annual high-water marks bore cacti and spiny desert shrubs widely spaced over the rock and sand. This was the essential desert, clean and elemental, in hues of red and tan. To look upward was to feel my sight and spirit soaring first to one massive wall and then to the other, until my spirit expanded to fill the whole vast reach of the canyon. If I lowered my eyes I saw cool, liquid green. The silent immobility of the desert above gave way to the constant motion of the river, quietly fluid in the deeps and slapping in noisy white waves over the bouldery shallows.

The wave slaps were wake-up calls in what might otherwise have been a too-dreamy progress down the great river. Each rapid announced itself with a distant grumble that grew to a roar and, if large enough, a point where the river dropped from view as we looked downstream. The raft guides liked to reconnoiter, sometimes by standing on the raft's pontoon in the quiet water above the rapids. That was the point where we hung momentarily poised, waiting on the brink, before the swiftly accelerating flow pulled the raft down a stomach-dropping slide and slammed it into the standing wave beyond. Water feels much harder than a liquid when you are flung into it. Things usually grew chaotic beyond the first wave with a succession of drenching, less-defined waves. Riding a raft with a professional guide to do the hard work provides an easy adrenaline rush. At the bow I felt like a nonchalant veteran. Then we reached the Inner Gorge.

The dark, steep, rugged walls of the Inner Gorge

The most spectacular canyon-within-a-canyon along the upstream half of the Grand Canyon, the Inner Gorge cuts into the ancient, extremely hard Vishnu Schist and Zoroaster Granite, leaving a deep, narrow channel riddled with huge boulders that form some of the worst rapids along the river. Even the entrance to the Inner Gorge is a visual metaphor for "Abandon hope, all ye who enter here." The dark rocks dip steeply upriver, creating the illusion that the river drops suddenly into the depths of the Earth. It was near this point that Powell allowed misgivings to creep into his usually ebullient journal, penning the lines that to me epitomize the uncertainties and just-below-the-surface fear of wilderness travel: "We have an unknown distance yet to run, an unknown river to explore. What falls there are, we know not; what rocks beset the channel, we know not. . . . Ah, well! We may conjecture many things. The men talk as cheerfully as ever; jests are bandied about freely this morning; but to me the cheer is somber and the jests are ghastly."

During my first river trip, my cockiness was effectively beaten out of me in a one-two blow at Sockdolager and Grapevine rapids. Powell's crew used nineteenth-century slang for "knockout blow" to name Sockdolager, and it gave them a worse time than it did me. Our descent began with a rush into a yawning hole that doubled the boat in on itself, squeezing me in the crease and sending my camera crashing into my head. I came up stunned and spluttering, but I was nineteen, and stubborn. The boatman told me he'd been watching me at the front of the boat through the whole trip, and "now it was time," presumably time to test my mettle. I remained at the front of the boat as everyone else moved back. Grapevine came up fast. I gripped the safety rope as tightly as I could. The first wave slammed me onto the floor of the raft, lacerating my hands on the rope. This time I came up gasping, my face probably as white as the wide-eyed faces at the back of the boat. I moved and spent the rest of the mercifully short day in the middle of the boat, shivering uncontrollably and dreading any more rapids.

Hikes up the side canyons provided a good respite when the Colorado's rapids left me chilled and wobbly legged. *Side canyon* seems a prosaic term for these worlds within a world, but then they are often prosaically named on the topographic maps. The maps give few clues to the existence of these secret pockets of cool green with clear water running quietly through them. The maps show only deep tributary canyons entering the main canyon.

Buck Farm Canyon was my first hint that names did not do justice to the side canyons. I hiked up from the main channel little suspecting that the tributary might be different. The trail began on a steep, exposed slope where the air vibrated with heat and light. Then I crossed an invisible divide and descended into a canyon cool and dark with the shade of fat-leafed cottonwoods, where starbursts of red and yellow columbines shot forth from rock walls cushioned by ferns and moss. A crystalline brook flowed through the canyon, falling over rock ledges into pools that made me feel languid after the icy slaps of the Colorado's rapids. Ouzels bobbed up and down on the edge of the plunge pools. Snake trails traced lazy undulations in the dust. The clear notes of a canyon wren reverberated in a waterfall of sound from an unseen perch high above, and the distant Colorado made only a subdued roar.

Each side canyon has its own character. At the Little Colorado River, milky turquoise-blue water forms a striking contrast with the red-and-tan rock walls and the royal blue sky. Cream-colored

Waterfall in Saddle Canyon, a side canyon within the Grand Canyon, Arizona

sediment the consistency of yogurt makes for sticky footing in the shallows, but the center of the stream is a fast current cascading over travertine steps. The travertine is calcium carbonate dissolved from the massive limestone layers that make up much of the canyon. The amount of carbonate carried in solution is delicately balanced with the water's temperature and dissolved oxygen. Any drop over a step or ledge creates turbulence and aerates the water, causing some of the carbonate to precipitate on the streambed. With time this material builds the travertine ledges and terraces that give many of the Grand Canyon tributaries their distinctive appearance.

The dissolving action of water on limestone creates intriguing mini-landscapes in the Grand Canyon. The Redwall limestone is honeycombed with caves large and small, and with tunnels where water flows in subterranean rivers. At Vasey's Paradise, water fountains up from a seemingly solid wall of Redwall limestone, forming a garden of ferns and mosses that Powell named for botanist George Vasey.

Some of the Redwall caves have sheltered both people and animals from the desert sun and provided a defensible home. At Nankoweap Creek, a cave high up the canyon wall holds the remains of stone-walled granaries built there by the Anasazi Indians generations ago. As I perched on a steep rock wall outside the granary, the narrow landscape of the canyon bottom expanded to a long, vividly colored riverscape. The rock immediately over my head formed a high red arch. Beyond this lay an immense curve of red and tan formed by the cliff across the river. The Colorado flowed pale green below, and an intensely blue sky arched over everything. I thought of the Anasazi. How did they perceive the world, knowing only this canyon country of vertical rock walls, desert flats, and then the big river flowing through it? Much as I loved the riverscape, I felt a lifting of my spirits whenever I climbed to one of the broad desert plateaus within the greater canyon. Did the Anasazi love the wide spaces, too, or did they feel threatened by the vastness? Earlier that day we had floated past an Anasazi bridge of ponderosa logs spanning a gap high in the rock walls. They must have explored the region thoroughly over a period of generations. I wondered, did they have their Powells, or was the exploration gradual and communal? Did their young people perform feats of exploration in lieu of vision quests? We can reconstruct so little of a culture from nonwritten physical remnants, but I found it impossible not to feel a kinship for the ancient ones as I watched the shadows of late afternoon grow deeper and felt my body relax.

Relaxation is one of the most precious gifts the river gives, a gift that I appreciate more as I age. On my first trip, I gained the most profound sense of place and time just by lying motionless on the beach as dusk fell and the stars appeared. Some nights we camped by a rapids, and my dreams were filled with motion and struggle. Other nights I spread my bag on sand still warm with the day's heat, a grove of tamarisks or a cluster of boulders forming my headboard. I watched swallows slipping between the edges of the breezes. Bats replaced the swallows as the light faded, skittering back and forth against the pastel colors of the evening sky.

I did not come to such relaxation immediately. I have been on camping and hiking trips since I was a few months old, but camping always meant a tent. A tent comes between you and the stars, but it also comes between you and rain, wind, insects, inquisitive wild animals, and whatever else you might care to ignore while sleeping. We had a very limited luggage allowance on the Grand Canyon trip, and I decided to take only a tarp that I would tie at an angle over my bed, like a lean-to. My wilderness skills did not encompass efficient tarp use, so the first challenge was to keep the tarp from hanging into my face while I slept. Our first night out I got slowly into my bag, reluctant to leave the sight of stars magnificently bright and plentiful. I got comfortable and was just dropping off to sleep when something dropped off the tarp onto the pillow beside me with a loud plop. It began immediately to crawl into my sleeping bag. I grabbed it and threw it outside, and then began to explore with my flashlight. I never found it, but it felt like a toad. Another night, after I had graduated to sleeping on top of the tarp, I had just started to fall asleep when I was awakened by a loud crackling noise from the tarp next to my ear. I rolled swiftly away and grabbed my flashlight, imagining a bighorn sheep about to step on my head. The flashlight's beam illuminated a small, startled frog.

I began sleeping on top of the tarp the third night of the trip. We'd had a strong headwind all day, and I was alternately choked by the neck strap on my hat or pelted by little water-bullets. I'd developed a headache by the time we reached camp in an unusually wide portion of the canyon and was not willing to put much effort into setting up my tarp. I fell asleep with the tarp hanging down into my face, but I was too tired to care. At some point I awoke, my headache gone, and simply pulled the tarp away without getting up. The effect was like unveiling a black velvet cloth thick with sparkling jewels. I had never seen a sky like that. The Milky Way formed a thick creamy swath, each star so large and clear that it seemed I could reach up and touch it. Every few seconds a shooting star made a brief light trail across the sky. I felt a peacefulness and unity with the universe I had never before experienced. The sky above the canyon walls formed an inverted bowl spilling out stars and dazzling me with stardust. I felt cradled and protected, moving effortlessly through space with the planet.

That was a turning point in the trip, an epiphany that I spoke of to no one at the time. The next evening, I walked slowly down the beach, thinking about the age of the rocks, the power of the river, and my own insignificance. What insight I have gained from wilderness has come at moments such as that star-struck night, when I have unexpectedly realized that all my energy and work and dreams are as nothing before the ageless reality of the natural world. I am calmed and steadied by understanding that it is good to work and strive, but that beyond all that is something more, which both contains and supersedes me.

Waking the next morning, with sand rubbed into my scalp and the colors of the canyon flat in the pale dawn light, I stumbled down to the river and shocked myself awake in its numbing cold. The night's adventures of packrat and mouse had left the sand tracked, and I could read in it all the tragedies and triumphs of another world. Between my bed and the river's edge, I passed through three pockets of alternately warm and cold air. By the time I had finished my morning's exploratory hike, they were all filled with scents of breakfast cooking and another river day well begun.

I felt like a real "river rat." Despite my morning wash, I was grimy and unkempt, and I didn't care. I was so tan that I looked dirty even after washing. I was like the members of Powell's 1871 expedition, whom Dellenbaugh described as "a shaggy-looking lot." Blisters bubbled my toes, scratches etched my legs, scrapes covered my knees—souvenirs of the side canyons. My gear was sandy and water stained. Repeated wettings and dryings had left my pants so stretched that when they were wet I had to hold them up to keep them on. I ate every meal as though a famine was imminent, in keeping with the tradition of "float and bloat" raft trips. The constant roar of the river was so much a part of my consciousness that I noticed the sound only by its absence in the side canyons. My legs had adjusted to the slipping-pumping motion of walking on sand, and when I lay down at night I still felt the world rocking around me. I was steeped in the sounds and textures of the river.

I loved waking up on the river and looking down to the rafts tugging at their ropes as though eager to be off and following the current onward. In the pulse and throb of the whitewater, the whole vast world condensed down into a momentary silence and then exploded outward again in chaos. I loved the utter slothfulness of flat water in the afternoon sun, when we forgot to wonder what lay beyond the canyon walls. And I loved equally the secrets of the side canyons, where shadows and ferns made the Colorado only a memory of brightness and heat. I watched the morning sun warm the cliff tops to bright colors and evening sun inflame them to incandescence. Writing in my journal, I had only to glance up to see a single, white-blossomed sand verbena surrounded by wind-rippled sand, or the Big Dipper framed in the narrow patch of sky between the canyon walls. Most of all, I loved simply being there to see for myself, each experience sinking into me slowly.

The raft trip ended much too quickly. We had a closing ceremony on the last night, with jokes and skits, and the presentation of certificates signed by the right hand of Powell, the hand he lost at the Battle of Shiloh. I hated to leave the river and to lose the camaraderie developed during the trip. We hiked up the Bright Angel Trail, every step taking us away from the private world of the river and back to the very public world of the much-visited South Rim. I sat on the rim gazing back down into the canyon, marveling at the wonderful hidden mysteries of the river that seemed unimagined and unimaginable from the top of the canyon. People around me looked briefly over the rim, took photos, made jokes, went away. They seemed to me irreverent before the altar.

I found it hard to sleep in my bed that night, confined by the walls and ceiling of my room. The river trip had given me lessons in humility and awe, but it had also awakened an even stronger love and feeling of protectiveness toward the Grand Canyon. Most of all, the river trip kindled a strong desire for more river trips, lighting a fire that I have not yet quenched more than thirty years later.

River Days

Paradise Lost

The great myth of the West is that there has been a winning of it. . . .
We had uncovered instead layer after layer of loss.
Douglas Preston, *Cities of Gold*

Two years later I returned to the Colorado River, hiking down
to Phantom Ranch to meet a raft trip on its way downstream to
Diamond Creek, the usual takeout point for Grand Canyon float
trips. Once on the river, I quickly slipped back into familiar patterns: wake before dawn and move about freely; hide your flesh
beneath clothing as the sun grows higher in the sky; brace and
relax, brace and relax as we pass from rapids to swift, smooth
runs. We camped early that first day, for the Colorado was rushing us downstream on a flood.

The river normally flows between twelve thousand and eighteen thousand cubic feet per second through the Grand Canyon,
but this was May 1984. An El Niño circulation system had brought
record snowpacks to the Colorado Rockies the preceding winter
and the winter before that. Now all the mountain tributaries of

the Colorado River swelled with snowmelt, together building to a tremendous volume of water that surged downstream toward the series of dams that stopper the mighty river along its course. Engineers at Glen Canyon Dam, immediately upstream from the Grand Canyon, were again nervously watching the water level in Lake Powell rise. The near-disaster of the preceding year was fresh in their minds and influenced their response to the 1984 high flows.

The dam operators had not anticipated the extreme snowmelt of 1983. By the time they had realized the amount of water they needed to let pass by the dam, the reservoir level was already dangerously high. They opened every outlet from the reservoir to release as much water as they safely could. At the peak flow, more than ninety-seven thousand cubic feet per second passed alongside the dam. The huge concrete spillways that burrowed through the reddish orange sandstone walls beside the dam spurted enormous white plumes of foamy water. Then the spillways began to spurt chunks of concrete the size of Volkswagens. The water roaring down the spillways moved with such force that even tiny joints in the concrete lining—similar to the cracks between blocks along a concrete sidewalk—formed an irregularity sufficient to start cavitation. Cavitation occurs when subtle pressure changes in fast-flowing water allow air bubbles to alternately form and then collapse. Each collapse sends out minute shock waves, and these waves weaken nearby surfaces like miniature jackhammers breaking concrete. As the shock waves broke apart the concrete lining of the spillways, the surface irregularities grew larger and the rate of cavitation increased. Soon the water eroded through the eleven-foot-thick concrete lining and began to dig a stairlike pattern of huge holes into the underlying sandstone.

The engineers watched the reservoir level continue to rise. They bought all the sheets of plywood they could find in the nearby town of Page, Arizona, and placed these incongruously fragile-looking splashboards at the top of the enormous spillway gates to keep the water from overtopping the structure and compromising the whole dam. As the damage to the spillways grew worse, the engineers had to shut the flow off lest the erosion work its way into the power generators. None too soon, the rate of snowmelt slowed and the dam held.

The next year saw even more snowmelt. The spillways had not yet been fixed from the 1983 damage, so only the generators and bypass tubes could be used to send water past the dam. These were all kept fully open. The reservoir level never reached the 1983 high point, but those of us floating downstream had an exciting ride.

We were rushed along on flows of 50,000 cubic feet per second. That would not have been an unusually large spring flood in decades past. Measurements of flow on the Colorado began in 1921 when the U.S. Geological Survey placed a stream-gauging station at Lees Ferry. The annual spring flood averaged approximately 86,000 cubic feet per second during the next forty years, reaching as high as 225,000 cubic feet per second in 1921.

Geologic evidence indicates even higher flows before stream gauging began. Tributaries of the Colorado in the Grand Canyon are mostly small, steep drainages. When a summer thunderstorm brings brief but intense rains to these small catchments, there is little vegetation to hold sediments in place. The thundershowers strip sand and boulders from the unprotected hillslopes and carry the sediment down the tributary channels as flash floods and debris flows. These flows gather enough momentum to continue down to the Colorado River, where they form debris fans that constrict the river and create rapids. When a flood comes down the Colorado River, the flow through the constriction is so forceful that it erodes the debris fan until the Colorado's path is sufficiently widened.

Flows on the tributary channels periodically rebuild the constrictions along the Colorado. Any particular tributary may have a flash flood or debris flow once every twenty years or once every hundred years. And the Colorado periodically widens the constrictions whenever it has a sufficiently large flood, approximately once every couple of hundred years. In the course of this seesaw-like change along the Colorado, most of the tributary constrictions attain a stable form. Geologist Susan Kieffer was able to estimate from these constrictions that the maximum flood along the Colorado River during the past few hundred years has been approximately 400,000 cubic feet per second.

Other features along the Colorado River also suggest huge floods. Rising floodwaters along the main channel backflood the mouths of tributary canyons, as well as the limestone caves and alcoves along the river. In these protected areas the floodwaters slow and the silt and sand suspended in the water gradually settles out. These sediments are left behind as the floodwaters recede. The sediments may remain in place for thousands of years in a dry, sheltered spot, each large flood adding its own increment to the deposit. The top of each sediment increment provides a minimum high-water level, and bits of leaves and twigs mixed in with the sediment can be analyzed with radioisotope methods to estimate the age of the flood. Using such sedimentary records, I worked with my friends Jim O'Connor and Lisa Ely to develop a record of Colorado River floods extending back forty-five hundred years. We found that the largest discharges during this time span were approximately 500,000 cubic feet per second, with ten floods exceeding 240,000 cubic feet per second during the last two thousand years.

Even the more moderate annual floods of 86,000 cubic feet per second recorded by the Lees Ferry gauge no longer occur regularly along the Colorado River. Beginning with Hoover Dam, completed in 1936, the Bureau of Reclamation industriously plugged the Colorado with more than six large dams, including Flaming Gorge in 1962 and Glen Canyon in 1963, both upstream from the Grand Canyon. The annual peak flood in the Grand Canyon was reduced to about 25,000 cubic feet per second. The bureau has come in for a tremendous amount of criticism as environmentalists have realized the devastating effect of these dams on the ecology of the Colorado River basin. But during the heyday of large-dam construction in the United States—from the 1930s to the 1970s—the dams were perceived as a clean source of energy and as insurance against the floods and droughts that played havoc with the regional economies of the western United States.

I began to understand some of the effects of the Colorado River dams during my second float trip through the Grand Canyon. Initially the high water and rumors of impending dam failure upstream only added excitement to the trip. Crystal Rapids was so dangerous given the higher flow that our raft guide made us walk around it, then flipped the raft trying to run the rapids as we watched. As other rafts came downstream, some of them flipped as well. No one was injured, and we regrouped downstream, flipping the rafts right-side-up, collecting floating gear, and providing reassuring hugs to those who had swum. One of the concerns related to the operation of Glen Canyon Dam is that, because the big floods no longer occur on the Colorado River, the precarious seesaw between tributary constrictions and main channel floods will tip toward the constrictions, and the rapids will become more dangerous. A pioneering 1923 survey of the streambed along the length of the Grand Canyon was repeated in 2000. The repeat survey showed substantial accumulation of boulder-size sediment along the river.

I fell asleep early the first night of the float trip, tired by my hike into the canyon and by the day's excitement. I awoke at first light and lay watching the river flow endlessly on in quiet boils between the dark canyon walls. I found it difficult to imagine the river tamed by dams. Better to have something wild and strong, beyond human control, to respect, if only to slow us in our thoughtless destruction of what we seemingly can control. I was comforted by the knowledge that the river had flowed by that spot throughout the night, unaided by human efforts or consciousness. The Colorado had flowed endlessly onward for thousands of years, and would continue to do so regardless of my fate. No river is permanent at geological time spans, but the Colorado provides permanence during a human time span. In that thought I find the reassurance of stability.

I brushed tamarisk flowers off my sleeping bag as I packed for the day. The tiny flowers form a faint lavender mist over the trees. The trees themselves grow in dense thickets along the waterline, providing shade and cover in the desert. But those who know the canyon's history are less impressed by the beauty of *Tamarix chinensis,* introduced to the American Southwest from Asia as an ornamental tree. Tamarisk was planted to form windbreaks in California's Imperial Valley and along many an irrigation ditch elsewhere. It thrived. Unlike many of the native riverside trees, such as cottonwoods or willows, tamarisk can survive dry spells when underground water levels drop. The seedlings cling tenaciously to moist sandbars and are not removed as easily during floods as the seedlings of the natives.

Within a few decades, tamarisk was crowding out the natives and spreading along river channels throughout the Southwest at a rate of twelve miles a year. The dense stands of tamarisk do not provide good habitat for many species of birds and other riverside animals that have evolved with the native tree species. The tamarisk stands are so densely intergrown that, when higher stream flows overtop the stream banks, the tamarisk thickets slow the water and allow sediment to be deposited. This sediment gradually accumulates, providing new germination sites for tamarisk seedlings, until the whole channel grows narrower and deeper. One of the Colorado's major tributaries, the Green River, lost more than half its width along some stream reaches in Canyonlands National Park during the 1950s as tamarisk invaded.

In the Grand Canyon, the lines of vegetation tell the story of historical river change. Tamarisks grow thickly along the contemporary high-flow level. Slightly above and behind the tamarisk thickets are the cottonwoods. These stately elders form isolated groves along the canyon bottom. They become more isolated with time, for the lack of large floods and the competition from the tamarisk prevent new cottonwoods from germinating. Forming a discontinuous line well above the tamarisk and cottonwoods are the hackberry trees that germinated during high floods in decades past. These too are no longer being replaced with new seedlings. Intermingled with the tamarisk along some portions of the river are new wetlands created by the relatively stable river flows. Because these wetlands shelter species at risk in the Southwest—Southwestern willow flycatchers, Kanab amber snails, and leopard frogs—the discussion of whether to restore historic high flows in the Grand Canyon is complicated by the knowledge that, although such flows might restore habitat for some species at risk, they would also destroy the new wetland habitat of other species.

While I studied the vegetation, the rising sun flooded the beach with light and the early morning coolness vanished quickly. We continued down through the Inner Gorge, the raft guides pointing out historical sites, birds, and anything else of interest that they spotted. Good raft guides are a pleasure to watch in action. Both the men and the women have superbly muscled bodies, and they move with unstudied ease and grace. They are companionable and easy to be around, they exude an aura of responsibility and competence, and they are knowledgeable about all aspects of the river. I have encountered some of them while away from the river, and been surprised at how diminished they seem. But on the river they strike me as omniscient, so that I find it hard to imagine them otherwise.

We entered the golden middle of the trip. Each trip has its initial adjustment period, when I am not yet at ease or accustomed to the new patterns of existence, perhaps comparing the trip to others I have taken. And each trip has its final phase, when I am already thinking of returning home and of what I will do next. In between lies the golden middle. This is the period when, on a good trip, I surrender my thoughts fully to the experiences of the moment. In the Grand Canyon, the world beyond ceases to exist. My thoughts are bounded by the canyon walls, and time is the flow of the river. Even the smallest new experience becomes a fully realized discovery. After the trip, I collect the experiences in my mind like snapshots in a photo album.

Shinumo Creek: Swimming through achingly cold water as I ascend between the undulating walls that the creek has carved into the black gneiss of the Inner Gorge.

Elves' Chasm: Stepped falls of rocks emerald bright with moss and ferns where the water combs itself out into sprays and plumes as it passes over the lip of each falls. Air sweet with flower perfume in the chasm, then heavy with water mist behind the falls.

Thunder River Falls: The powerful torrent of white water bursting from the arid red-and-tan cliffs. My hike to Surprise Valley above the falls, where I watch a collared lizard grab a smaller whiptail lizard. The whiptail is still alive and wriggling, but the collared gets the whiptail's head pointed into his mouth and proceeds to gradually gulp the whiptail down whole. Once the whiptail's head is engulfed, the body goes passive. As the collared works his jaws down his only slightly smaller prey, his body goes from slack to distended. It is as if I could eat a Newfoundland or a Great Pyrenees dog whole, without using my hands to ingest the animal. I leave the collared lizard panting from his exertions.

Deer Creek: Water twisting and turning far down the narrow slickrock canyon while I follow the winding of the rock myself, holding my breath, hands pressed to the rock wall as I squeeze along narrow stretches of the trail perched far above the water. Handprints precede and follow me where Native Americans too held their breath slightly, turning sideways to keep the precarious trail. Back down again, from red rocks to green water, I watch ouzels slipping into the flow and swimming about in the turbulent depths, swift and graceful as penguins. At the canyon's mouth, water plunges over a falls, hammering down onto the anvil of a broad pool. I wade backward into the pool, leaning hard into the powerful current and spray-laden wind until I can go no further, taut between my own force and the force of the water.

And Havasu Canyon: The raft guides have nicknamed it "Havazoo" because every raft trip stops at this most lovely of side canyons. I am lucky. Early one dewy morning my boat is the first to reach Havasu that day, and I am first off the boat. Hardly into the canyon, hiking alone, I see a bighorn sheep. There follows one of those moments that are wonderful precisely because they are unexpected. I suck my breath in with surprise and remain motionless. Seconds stretch into minutes while the powerful ram makes his leisurely way along the cliff. With his unconcern at my presence, he seems the epitome of a wild animal. I continue alone, but not alone, into the silent, shadowed canyon.

Edward Abbey wrote that Earth is the only paradise we ever need. Havasu Canyon on that morning was paradise. I remember it now as colors. Emerald green water flowing over travertine terraces beneath the forest-green trees. Yellow acacia blossoms sweetening the air. Leaves and water playing with the sunlight streaming into the canyon, tossing the light back and forth until it is rarefied to palest greens of jade and epidote. Where the canyon floor stair-steps over falls, the white foam of the water plunges over brown travertine draped over rock steps, continues into azurite-blue pools, then quietly spills off the sides in aquamarine terraces and delicate steps. Pools succeed falls, as masterfully arranged as a Japanese garden. Even the water feels soothingly warm.

The side canyons left calm, contemplative photographs in my mental album. The rapids left images that better resemble videos with slightly blurry, hurriedly taken sweeping shots and momentary freeze-frames of startling clarity. Lava Falls is the

Deer Creek, a tributary of the Colorado River in the Grand Canyon

culmination of the Grand Canyon's big water. Rated ten-plus, or not recommended, on the canyon's one-to-ten scale of rapids, Lava Falls drops over basalt ledges that constrict the Colorado. Powell, seeing the brown water dash itself into a frenzy against the black rocks, described in his account of his river trip "a river of molten rock running down into a river of melted snow." I studied these rapids of fearsome reputation from an overlooking cliff while our raft guides reconnoitered the route. I was disappointed. The rapids looked relatively short and not all they were cracked up to be. Approaching them from water level was another story. I experienced a moment of sheer amazement as I realized the true size of the waves. Then the waves were crashing all around us and into us, the boat rocking and heaving madly. In an instant we were in water up to our knees and I was alternately bailing and hanging on for all I was worth, all through that long "short" run. Safely anchored at a beach downstream, we saluted the raft guides with river-chilled beers, feeling springs of joy at our own daring and survival bubbling up out of us. I found that I was still wobbly-legged two hours later as we hiked up to see Anasazi petroglyphs above the river.

The river corridor is rich with archeological sites. Hunter-gatherers passed through the region some four thousand years ago. Anasazi farmers settled the old, high riverbanks circa A.D. 700, remaining there for more than four hundred years. The turbid waters of the big floods used to deposit silt and sand well above the low water level, and these higher benches made good farming sites. The closing of Glen Canyon Dam reduced flood levels and trapped much of the fine sediment that periodically rebuilt the benches. Now rainfall erodes the benches, exposing archeological materials and gradually washing them down to the

Lava Falls Rapids

river. And the clear, erosive waters of the river itself eat away at the bases of the benches, undermining them until they collapse. Lower sand beaches all along the Grand Canyon have grown dramatically smaller since the closing of Glen Canyon Dam, reducing both habitat for riverside trees and animals and camping sites for rafters traveling through the canyon. The side canyons still bring sediment to the Colorado, but there remains less sand and silt than the river carried historically. Perhaps more important, much of the sediment may be stored among the huge boulders on the river bottom. Only floods the size of the large, pre-dam floods would be capable of stirring up this sediment and replacing it high along the river margins.

Processes of change in the Grand Canyon seem to play with our sense of time. There are no easy generalizations. Expedition photographer Jack Hillers created an extensive record of Powell's second river trip in 1871. Hal Stephens and Gene Shoemaker began to replicate these photographs in 1968, five years after Glen Canyon Dam closed. Geomorphologist Bob Webb has made many trips through the region since, replicating both the 1871 and 1968 photographs. This visual record shows arid canyon walls with so few changes that, in repeated photographs, it is possible to pick out the same long-lived desert shrub or large boulder. The record also shows the spreading sea of tamarisk crowding the river at Lees Ferry, and the changes in rapid configuration where a big debris flow has enlarged a tributary fan. Photos also record amazing alterations along the upstream portions of tributaries. Upper Kanab Creek, for example, has changed from a shallow, grassy swale to a gully cut seventy-five feet deep, then subsequently completely filled again in fifty years.

What the photographs do not record are changes in the canyon's wildlife driven by changes in habitat and river processes. Many years after my second raft trip, I served on a scientific panel reviewing the research conducted in the Grand Canyon by geomorphologists funded through a Bureau of Reclamation research center. As part of that review, I had the opportunity to snorkel along the Colorado River just below Glen Canyon Dam. The water released from the base of the dam is crystal clear and so cold that it chilled me through my dry suit. The main channel resembled an aquarium full of huge trout. Fish biologists stocked trout not native to this part of the Colorado River system throughout the Grand Canyon during the 1920s and 1930s. These introduced fish now thrive on water released from the base of the reservoir that averages forty-eight degrees Fahrenheit year round. As I swam downstream with the trout, I saw a small, shallow channel along the river's margin, into which I swam to warm up. I found a school of young humpback chub *(Gila cypha)* in the brightly lit warm water.

Humpback chub are one of the species native to the Colorado in the Grand Canyon. These fish evolved to live in the sediment-thick waters of a river that historically reached seventy-five degrees Fahrenheit during the heat of the summer. To provide nursery habitat for the young, they rely on backwater channels such as the one I had entered, but these channels are gradually disappearing along the Colorado. The chub also evolved to time their life cycles around the yearly gradual rise and fall of the Colorado. Now the river jerks up and down spasmodically on a twenty-four-hour cycle governed by the demands for hydroelectric power in Las Vegas. Flow through the Grand Canyon may rise from eight thousand cubic feet per second to twenty-eight thousand cubic feet per second during a single summer day. The cold clear waters, the increasingly simplified channel lacking in backwaters, and competition from introduced trout are all driving the chub and other native species to extinction. A 1996 census estimated eight thousand chub in the Grand Canyon. By 2001 the number had fallen to two thousand fish. A species that has been present in the Colorado River system for more than two million years, and is found nowhere else on Earth, is at the threshold of survival in a national park that many people consider a "living museum" or a safe preserve for a canyonlands ecosystem.

Sometimes changes in the landscape of the Grand Canyon region are so rapid and dramatic as to be readily perceived. I visited Lake Powell during the spring of 2005, when water levels were at the lowest point since the construction of Glen Canyon Dam in 1963. Lake Powell was designed to serve partly as a buffer to insure that senior water rights in Arizona and California can continue to be filled during periods of drought. Water storage in the lake also insures that states in the upper Colorado River basin can continue to divert water from the river during times of drought. In other words, I can still drink Colorado River water in my home on the eastern side of the Rockies even if regional snow-pack and rainfall levels are very low. The drought that began the twenty-first century in the Colorado River basin has been testing the buffering capacity of Lake Powell. The lake normally has a live storage capacity of approximately twenty-five million acre-feet. *Live storage* refers to the portion of the lake basin not occupied by sediment, and an acre-foot is a unit of measurement equivalent to one acre covered by water one foot deep. By spring of 2005, Lake Powell was down to eight million acre-feet of storage.

The rapidly dropping water level left thick white "bathtub rings" of salts on the nearly vertical rock walls. Hundreds of logs weathered to a pale shade of gray lay piled on rock ledges where the receding waters had paused a moment. Upstream portions of the lake had once again become a river flowing muddy between extensive flats of newly exposed sediment that had been deposited in a delta at the head of the lake over the preceding forty years.

The surrounding terrain above the lakeshores looked strange enough. Gray hillslopes barren of vegetation and deeply cut by channels formed a badlands across which active sand dunes migrated on the plateaus below the snow-covered Henry Mountains. Below the badlands a contorted topography formed on the exposed lake sediments. Thousands of small tension cracks criss-crossed sediments slumping toward the river channel. White salts precipitated from the evaporating water crusted the sediment, but invasive tamarisk and Russian thistle nonetheless spread quickly in lines and patches of green. Shallow circular pools or the craters of mud volcanoes indicated spots where methane—released from decaying organic matter buried with the sediments—now bubbled up.

Walking carefully across the terra nonfirma of bubbling methane pools, I found it easier to conceive of the sixty-six million tons of sediment that the Colorado River had brought here each year during the forty years since the dam was closed. Some of this sediment is contaminated by naturally occurring elements such as selenium, which was leached and concentrated when excess irrigation water from farm fields drained back into the river. Some sediment includes uranium tailings dumped at the mouths of tributary canyons before the reservoir was constructed. Other portions of the sediment are contaminated with heavy metals, including cadmium leached from used batteries once dumped from the marinas along the lake. All these toxins, as well as the massive volumes of sediment accumulated in the upper reaches of the lake, are now being mobilized and carried down toward the dam as declining water levels allow rivers to once more erode delta sediments. I think of an interpretive display about Hite City, a short-lived settlement along the portion of the Colorado River submerged by Lake Powell. The display states that, "finally in 1964, the rising waters of Lake Powell engulfed Hite, ending forever what dreams of prosperity had begun." *Forever* may be too strong a word for a dam and reservoir with a very finite span of existence.

Lake Powell has fulfilled its intended purpose of helping to buffer lower- and upper-basin states against drought-induced shortages of water from the Colorado River. But Powell and other water storage structures throughout the Colorado River basin can only do so much. Tree-rings from conifers growing in the southwestern United States record droughts during the past few centuries that lasted much longer than the drought currently causing such consternation among water managers. Predicted temperature increases that will reduce mountain snowpacks and likely reduce rainfall, and rapid increases in population and consumptive demand in regions as diverse as the Colorado Front Range and southern and central Arizona, which draw on water from the Colorado River, provide further sources of concern about the long-term sustainability of water supplies in the Southwest. The dropping water levels of Lake Powell foreshadow more extensive landscape changes likely to occur in the Southwest as water becomes increasingly scarce in an already dry region.

.   .   .

My second raft trip was the beginning of reeducating myself beyond my first perceptions of the Grand Canyon as a wilderness paradise preserved from the influence of industrial-age humans. I floated down the whole river a third time many years later; I conducted research on prehistoric floods; I studied the tributary Paria River; I served on the review panel mentioned earlier; and I read much. I floated down the canyons of the upper Colorado River watershed: Gray, Desolation, Cataract, and Lodore on the Green River; the Yampa River through Dinosaur National Monument; and Westwater Canyon on the Colorado. I visited the Colorado River delta at the Sea of Cortez. This region, which Aldo Leopold described in 1920 as a lush forest and marshland laced with smaller stream channels and alive with deer and jaguar, is now largely a treeless, salt-rimmed mudflat. In most years the Colorado River no longer flows to the sea. It is siphoned off along its way by burgeoning cities, from Denver to Las Vegas, Phoenix, and Los Angeles;

dispersed across thousands of acres of water-hungry crops in the United States; and finally diverted into Mexican cities and croplands. The mighty river of the southwestern deserts is no longer a river so much as a canal feeding the water where and when humans choose. A good deal of what I find tragic about these changes is the wastefulness with which we use water and energy. Do we really have to generate hydroelectricity at Glen Canyon for a region with a superabundance of sunshine? Do we really need green lawns and flood-irrigated crops in the desert if our water-storage reservoirs drive other species to extinction?

On the last morning of my second river trip, I sat sipping coffee and watching the rising sun illuminate the canyon walls. Across the river, a band of bighorn sheep went slowly by. All around me the sand was busy with mouse tracks, as though they had held a square dance round my head all night. I was reluctant to return home. I felt as though the trip had returned the sense of joy to my life, and I remembered Whitman's words from *Leaves of Grass:*

> Now I see the secret of making the best persons,
> It is to grow in the open air and to eat and sleep with the earth.

Each time I return to the canyon country, I feel again this precious sense of peace and belonging. Now it is inextricably mixed with a sense of loss and chagrin at the changes my people have wrought in this incredibly beautiful and intricate landscape. These feelings of loss foster a sense of responsibility. Because I know what has been and is being lost, I feel an urgency to communicate that knowledge to others. My only hope is that knowledge, once sufficiently widespread, will foster in all of us a sense of responsibility and a resolve to modify our society's destructive patterns of resource use.

The Delicate Strength of Rock

I have seen almost more beauty than I can bear.
Everett Ruess

The slickrock country of northern Arizona and southern Utah is one of my favorite spots on Earth. There the rivers have carved deep canyons into broad plateaus with euphonious Native American names: Paunsagunt, Markagunt, Kaiparowits, and Kaibito. Exuberant creeks run clear and cold on the plateaus, and on cool mornings the drumming of a woodpecker is loud as a jackhammer in the quiet before dawn. Aspen grow straight and slender in bowl-shaped depressions, the dark firs shadows among them. Aspen groves are bright and young hearted, glowing pale peridot green even under cloudy skies, as though the aspen leaves provided their own source of illumination. The leaves tap softly in the breeze, providing a rhythm for the melodies of birdsong all about. And when blue-gray dusk falls in a velvet curtain, the aspens show pale green long after the conifers are lost in the fading light.

Below the plateaus, pale fingers of sandstone stretch out into the flat, green pastoral valleys. Olive-green pinyon and juniper dust the layered buff-and-red cliffs edging the plateaus, and the wind rasps against pine needles and rock. The cliff faces are scarred and tracked, their history recorded in sweeping beds of sand left by ancient dunes and in checkerboard fractures streaked with the brown desert varnish left by a thousand rainstorms. The sinuous layers in the rock are so well preserved that I can almost hear the soft rustle of the wind sweeping sand grains over the ancient dune crests. The rock seems to be alive and flowing as the massive beds swerve across the vertical walls of the cliffs. The walls weather into great rounded beehives or piles of enormous discs, creating a vast sandstone sea with wave troughs and crests, swells and breakers. The rock is clean and vivid, only the barest decoration of pinyon or manzanita clinging to the crevices without benefit of soil. This is the Navajo's Land of the Sleeping Rainbow, a land of vertical rock unsubdued and as elemental as the bones of a skeleton.

For a person standing on a narrow finger of rock hundreds of feet above the adjacent valley floor, the world can condense to a tiny point of rock surrounded by space. Elsewhere it can drop down into slot canyons incandescent with light and color at midday. The slot canyons preserve the history of the canyonlands. Flood sediments tucked into caves and rock-roofed alcoves along the channel sides record ten thousand years of floods down these mostly dry channels, as well as repeated episodes in which the canyons filled with sediment and then emptied down to bedrock once more.

The same Paria River that can carry huge volumes of silt and sand down to the Colorado River above the Grand Canyon also cuts a twisting labyrinth of slot canyons into the sandstone mesas of southern Utah. In April, when the snow of the high-country slushes into mud-laden torrents, the Paria flows pale orange-brown. Early settlers might have described the Paria, like the Platte River in Colorado, as too thick to drink, but too thin to plow. The turbid water hides the streambed, and hiking through it can be perilous guesswork. But hiking is the only way to access these canyons, so hike we did.

The river begins as a sand-bed channel looping widely across a broad valley of sagebrush and low hills. When it crosses the thick rock layer that geologists call the Navajo Sandstone, the Paria begins to cut down rapidly, and less than an hour's hike downstream the river is deep in a slickrock canyon. Vertical walls of rust-red, orange, and buff sandstone reach upward to improbable heights. Hiking with my friend Sara Rathburn, I found it difficult to realize the scale of the walls as I looked up- and downstream, unless I could see a small human figure dwarfed by the cliffs. Silt marks left by old floods line the canyon like bathtub rings, and we were there to study the flood sediments.

As we continued downstream we detoured up tributary canyons such as Buckskin Canyon, where the undulating canyon walls shut out the sky and create a dusky gloom. In places the canyon becomes so narrow that we could touch both sides at once. I derive a sensuous pleasure from such sandstone walls. Smoothed by the abrasive sand grains carried in a thousand floods, the walls are sinuous as the tracks of a sidewinder. On that first trip they were a curiosity, inspiring our awe and a desire to take photographs.

Back in the main canyon of the Paria, we followed the river downstream as the height of the canyon walls slowly increased to a thousand feet. The river burrows into the Earth, leaving the rock formations behind like a strip chart record and slowly losing

Sinuous crossbeds in Navajo Sandstone

touch with the world beyond. We too went deeper into the Earth, losing touch with the outer world as we became engrossed in the flood sediments and in the daily pleasures of sun on rock or hot chocolate after supper.

When we finished the work, we continued downstream toward the junction of the Paria and the Colorado. We passed from the clean, massive walls of Navajo Sandstone into the horizontally layered siltstone, sandstone, and shale of the Kayenta and Moenave formations. These rocks formed a narrow inner canyon below the Navajo walls. Flutes and layers along the inner canyon created bedrock ledges that tempted us to climb out along them until they narrowed suddenly and we were trapped between a vertical overhanging cliff and the swift, cold water strewn with boulders.

The valley widened dramatically toward the end of that very long day. The Paria flowed through a sandy channel between broad, low plateaus sparsely vegetated by saltbush, prickly pear, yucca, and an occasional heat-blasted willow. Beyond the flats the rock walls towered at a distance: orange-red, massive sandstone underlain by pink, purple, and brown shales crumbling into house-size blocks. As the light faded we looked back across the valley we emerged from, a moonscape of angular talus beyond which a single shaft of sunlight illuminated a field of pale orange sand dunes surrounded by rock walls deep lavender under the stormy sky.

The work from that trip produced a record of prehistoric floods covering the past forty-two hundred years. Eleven sizable floods left sediments along the Paria River during that time span. The biggest of them was more than two and a half times larger than the biggest flood measured since stream gauges were installed in the early twentieth century.

Even more impressive than the record of floods was the evidence that the Paria River has cut eighty feet down into its sandstone canyon in only twenty thousand years. This pace of change is geologically fast. Yet part of the mesmeric quality of the landscape of the canyons and plateaus is the feeling of timelessness and permanence. This region, in which ancient peoples including the Anasazi and Fremont lived for thousands of years, was one of the last areas of the continental United States to be explored by European Americans. Their reactions varied from the prosaic to the reverent. Writer John Bezy quotes the Mormon pioneer Ebenezer Bryce, who summarized the Utah canyon named for him as "a hell of a place to lose a cow."

Subsequent visitors found themselves drawn repeatedly to the vivid colors and clean, massive lines of the sandstone cliffs. Arizona's territorial historian Sharlot Hall was one of the first writers to explore the terrain north of the Grand Canyon. She spent more than two months wandering the region on horseback in 1911 and wrote, in *Sharlot Hall on the Arizona Strip,* of "many a still unheralded natural wonder." The region became a spiritual center for writers Willa Cather, Ed Abbey, and Ann Zwinger, whose eloquence attracted others. And beyond the writing remains the landscape itself, enigmatic and achingly beautiful.

I found my mind returning to the sinuous canyon walls I photographed on the Paria trip. I wanted to understand how the walls had come to be shaped that way. I wanted to live among the canyons once more.

I returned to the canyons a few years later. By that time I was a professor at Colorado State University, and I brought my graduate student Susan Fuertsch with me. We camped at the head of Wire Pass Canyon, one of the main slot canyons that branched off the Paria. We arrived after dark, setting up camp near the sandy wash that abruptly funnels into the deep, narrow canyon. During the night it began to rain. There are people who cherish rainy mornings when camping, enjoying the patter of raindrops on the tent fabric and believing that such mornings invite leisurely rising. Susan is such a person. Then there are people who consider morning rain when camping, particularly in the desert and with work waiting, an affront. The rain disrupts the schedule and is contrary to the expected nature of things. I am that type of person. Susan tried hard to ignore the impatient rustling coming from my sleeping bag as gray light gradually illuminated the tent. But I hustled us both out to work as soon as the rain slowed.

The sky cleared as we followed Wire Pass downstream to its junction with equally narrow Buckskin Canyon, searching for sites at which to measure the canyon's geometry. Having scouted the short length of Wire Pass Canyon, we returned to the upstream end and spent the next few days surveying our way slowly back downstream once more. The weather quickly returned to the usual high-desert heat of late May. Even though some of the exposed sandstone uplands acted as solar ovens, Susan and I had to cover every inch of flesh against the tiny biting gnats that appear in the canyonlands in late spring and early summer.

When we finally worked our way back down to the junction of Wire Pass and Buckskin, I was amazed to find fresh silt lines three feet up the walls of Buckskin. Apparently the first night's rain had sent a flash flood down that canyon, while Wire Pass remained safely dry. We might easily have chosen to begin work at the downstream end of Wire Pass and been caught in the flash flood. Susan took ample advantage of this opportunity to remind me of the values of patience and caution.

I returned a couple of years later to survey Buckskin with another of my graduate students, Doug Thompson. By the time Doug joined me at the canyon, I had been camping and working alone for several days. He surprised me with half a gallon of soupy chocolate ice cream he had purchased at a little store on the drive in. The quantity was more than even two serious ice cream eaters could finish, and we put the half-empty carton out away from our tents. Later that night we heard a persistent rustling near the carton. A flashlight beam revealed a mouse gnawing at the desiccated mound of sugary chocolate. The mouse had reached heaven and blithely ignored us and our light. Presumably the ice cream tasted better than the hood lining of my car, which the mouse also consumed during my two-week stay.

Doug and I learned to work in the dry, exposed sections of the canyon early in the morning, saving the shadowed portions or the sections with water for midday. Although the canyon often seemed a silent, motionless landscape, we shared it with plenty of other creatures. Large, pale-blue crayfish shot nervously backward when we crossed the pools, and beaver-gnawed sticks collected in the eddies. In the drier sections, little rattlesnakes buzzed warnings when we disturbed their midday siesta in a shady crevice. Collared lizards fled in a scattering of sand. Hiking back to camp early in the evening, we sometimes saw a hawk circle steadily overhead or a jackrabbit burst from the low brush beside the trail.

I had begun the project in the slot canyons guessing that the sinuous expansions and contractions in the canyon walls corresponded with places where fractures weakened the rock and made it more susceptible to weathering. Our fieldwork detected no correlation between the fractures and the undulations of the walls. We did find that the wall irregularities mimicked the form of turbulent patterns in the rapidly flowing water, suggesting that small differences in the rate at which the energetic floodwaters eroded the canyon walls gave rise to the undulations. We also discovered that the irregularities reduced variations in the rate at which floods expended energy as they moved downstream between these deep, narrow canyons and the intervening wide, shallow, sandy stream reaches. The undulating walls thus helped to stabilize the entire stream system.

I made one more work trip to the area, this time to the nearby Escalante canyonlands in July, to show the area to Lisa Ely and Hema Achyuthan. Hema was a fellow geologist visiting Lisa from India. Lisa and I had hiked and worked together in remote canyons throughout the Southwest, but Hema had never camped before. She proved to be adventurous but deathly afraid of the little rattlesnakes common along the canyon. At one point, as Hema moved quickly past yet another rattler while I distracted it with a long stick, she firmly declared that this trip was a Historic Event, and that she was sure no Indian had ever hiked these canyons before. She may have been right.

Our third day out, we awoke to clouds. The clouds looked more and more threatening as we hiked down one of the slot canyons, and lightning began to flash around us. Lisa's dog, Nimbus, kept looking back at us as if to say, Do we really want to do this? The rain began just as we entered a particularly narrow stretch of the canyon. It came as intense, stinging drops that felt almost like hail. Scurrying along with our heads hunched down into our shoulders, we were quickly soaked. In less than ten minutes, small waterfalls of frothing white rushed down the red rock faces. I sometimes think of the Navajo Sandstone as being a sort of rock sponge, very porous and permeable and relatively soft. This is

Doug Thompson in Buckskin Canyon, Utah

true relative to other rocks, but the Navajo *is* still rock, and it sheds rainwater surprisingly fast. A frothy brown stream soon flowed down the channel, forcing us to hop from boulder to boulder. Lisa and I were starting to worry about getting through the narrow stretch of the canyon before a real flash flood occurred, and Nimbus was definitely questioning our leadership skills. But within twenty minutes the rain stopped, leaving glistening black trails on the slickrock.

I pondered the abrupt transformation in the landscape as we continued downstream. I had read technical papers on flash floods, studied the aftermath of such floods, and lectured on them. I thought I understood flash floods. I did understand them, and I could explain in great detail and with many precise technical terms what had happened during those twenty minutes. But I did not truly appreciate flash floods until that day. It suddenly struck me that I hardly knew the canyonlands at all. I had not begun to experience all their hours and seasons. Perhaps that is why so many people who live as I do, without direct and continuing contact with a natural landscape, have so little understanding and real appreciation for the natural world. We can all enjoy the scenic or recreational values of a landscape, but how many of us really understand the processes that create the landscape?

Having gotten more of a feel for flash floods, I was impressed, on a later trip, by the evidence I saw of flash floods and debris flows along Warm Springs Wash, a tributary of the Yampa River in Dinosaur National Monument. At that time, I took a commercial raft trip down the Yampa and hiked a short way up Warm Springs while the raft guides stopped to reconnoiter Warm Springs Rapids. The rapids form where Warm Springs Wash has repeatedly dumped boulders and sand into the river, constricting its flow against the opposite cliff. The most recent addition occurred on the night of June 10, 1965, the noise of which frightened the wits out of a group of Boy Scouts camped downstream. The 1965 flow brought about fifteen thousand tons of sediment into the river bottom, including boulders fifteen feet in diameter. As in the Grand Canyon, the Yampa rewidened its channel between tributary flash floods, and the rapids changed continually over time.

Unlike the Colorado River in the Grand Canyon, the Yampa River is not dammed. A dam is now proposed on the Yampa River upstream from Dinosaur, but at present the river is the last major undammed tributary of the Green River and, ultimately, the Colorado River. As such, it presents a unique opportunity to study river processes in the absence of human interference with flows. I asked my graduate student Lauren Hammack to study the dynamics of Warm Springs Rapids, and she spent a summer diligently measuring boulders and surveying both the Warm Springs Wash and the Yampa channel. She endured blistering hot days except when I visited, when it inevitably rained and threatened to send another debris flow down on our campsite.

Because it is undammed, the Yampa River harbors populations of fish species native to the Colorado River. Most of these native fish are now threatened or endangered as a result of changes in flow and habitat or the introduction of exotic species. The Yampa, the Green, and other rivers of the upper Colorado River basin each alternate in a downstream direction between narrowly confined canyons and open valleys where the rivers meander widely and flood seasonally across the broad bottomlands. When people of European descent started homesteading the upper river basin in the 1860s and 1870s, they relied on crops that needed extra water. As the water table dropped from irrigation demands, sloughs and backwaters along the desert and plateau rivers were drained, or cut off from the main channel and plowed. Backwater areas provide crucial nursery habitat where young native fish may rest and feed in the warm, shallow water, protected from larger fish predators. Historically, the native fish moved widely through the river system, migrating between spawning areas on gravel bars, nursery areas in the wide valley bottoms, and feeding areas dispersed throughout the river system. The loss of nursery habitat during the later nineteenth century was followed in the twentieth century by the construction of dams that blocked migration routes, flooded upstream habitat, and stopped the downstream movement of nutrients. A river pulses through time and space, flowing downstream and spreading periodically across its bottomlands, and the lives of its fish depend on these pulses.

The latest injury to the native fish came in the early 1960s, just before the closure of Flaming Gorge Dam, when government biologists used rotenone to poison the native species in the Green River so that the river could be stocked with the exotic rainbow trout beloved by anglers. The poison was dumped into the river only twenty-four miles upstream from Dinosaur National Monument. Although there was an attempt to detoxify the river using potassium permanganate fifteen miles upstream from Dinosaur, the poison spread into the monument. The poison hit the Colorado squawfish *(Ptychocheilus lucius)*, the razorback sucker *(Xyrauchen texanus)*, the humpback chub *(Gila cypha)*, and the bonytail chub *(Gila elegans)*, all of which are now listed as endangered or proposed for listing.

The endemic species are impressive fish. Although the squawfish, now renamed the Colorado pikeminnow, is a minnow, adults can be up to six feet long and weigh seventy-five pounds. They may live more than fifty years and migrate up to sixty miles along the river. But size and longevity have not been enough. Three-quarters of the native fish species in the historically warm and turbid Colorado River system are endemic, or unique to this one place. By 1982, 76 percent of the fifty-five fish species present in the upper basin were nonnative, mostly game fish such as bass in the reservoirs and trout in the rivers. These introduced fish compete for food and space with the natives and prey on the eggs, larvae, and juveniles of the native species. Having evolved in isolation, the endemic native fish lack the competitive abilities of exotics from species-rich areas, particularly when the natives are stressed by habitat loss.

Habitat loss in the Colorado River system comes in many guises. The native fish return to the gravel spawning bars that their ancestors used, but these bars are no longer suitable for spawning. The adult fish must be able to dig a shallow depression into the bar surface with their fins, and the eggs must not be buried by fine sediment that cuts off the oxygen supply brought by flowing water, or exposed by fluctuating water levels. The closure of Flaming Gorge Dam on the Green River largely halted spawning on downstream bars, perhaps because of abnormal flows, low water temperatures, or an increase in fine sediment on the bars.

Before we can effectively restore habitat or operate the dams so as to help the endemic fish, we must understand their habitat requirements. This is not simple. Fisheries biologists study the requirements of fish eggs, which may differ from those of fry, which in turn use areas of the river different from those used by adults. And the differences between unsuitable, barely suitable, and ideal habitat must be quantified, because every drop of water in the Colorado River basin is fought over by humans keen on having an endless supply of irrigation water, drinking water, and hydroelectric power.

I worked with my graduate student Ed Wick, who has a background in fisheries biology, to study what processes keep a spawning bar along the Green River just outside Dinosaur attractive to pikeminnow. We wanted to know just about everything: the source of the sediment forming the bar, the level of flow that deposits the gravel and cobbles, and the level of flow that winnows away the sand and silt, keeping the gravel clean for spawning. These were not "academic" questions. Ed's model of river processes maintaining the bar was used to request levels of flow release from Flaming Gorge Dam to preserve this valuable habitat. It seemed that everyone was looking over our shoulders. The water users do not want to release any more water than necessary. And the scientific and environmental communities know that the pikeminnow is running out of time.

Ed regularly surveyed the streambed, measuring flow velocity and sediment movement as flow levels changed. Collecting data, he cheerfully risked life and limb (never mind whose) by stretching a nylon cable across the channel during high flows and attaching a small boat to the cable in order to stabilize the boat while making measurements. On one of my visits we detached the boat from the cable in order to go upstream. We went only a short distance before deciding that we needed to tie flagging to the cable stretched about three feet above the water, in case any unsuspecting rafters came downstream. But as we attempted to tie on the flagging, the boat shifted downstream and the cable stretched taut around the outboard motor. Ed was at the back of

Delicate Arch, Arches National Park, Utah

the boat, and he unthinkingly pulled the taut cable over the top of the outboard motor and released the cable toward the front of the boat, where I sat, launching me out of the boat like a projectile from a slingshot as the cable caught me in the chest. I was holding my nonwaterproof camera, and so launched myself back into the boat almost immediately, no real harm done. My most vivid memory is of Ed's face. Undoubtedly he was thinking of his upcoming preliminary PhD examination.

Ed's research resulted in recommendations that the Bureau of Reclamation change water releases from Flaming Gorge Dam to mimic natural flow variability, with higher releases during wet years and more seasonal variability during both wet and dry years. In the absence of the political will to force the bureau to change its modus operandi, the recommendations have yet to be implemented.

．　．　．

The canyonlands provided me with invaluable lessons. When I ran out of water while hiking the Under-the-Rim Trail at Bryce Canyon National Park during one of my first long day-hikes in the West, I learned that force of will won't get you too far if your body is breaking down from dehydration. When my parents and I shared a dozen orange Creamsicles on a hot June day, devouring them with the zest of campers long removed from amenities, I collected a memory that I now cherish. When I first hiked the Kolob Trail from the uplands of Zion National Park down into Zion Canyon, I was humiliated at having to crawl along the thin rib of rock bridging space out to Angel's Landing. When I returned again, knowing what to expect and nerved for the challenge, I found it easy to cross the rock bridge. When I felt a flash flood churning about my legs, I perceived it in a way that years of study had not prepared me for. When I learned that native fish could no longer survive in the mighty rivers, I could hardly reconcile my new knowledge with the remote wilderness in which I had reveled at eighteen. When I learned that knowledge alone is not enough, I moved further toward political activism. And when I contemplate the delicate strength of rock stretched almost to the breaking point in a slender red arch defying gravity against the blue sky, I realize that all my measuring and modeling and thinking will not by themselves bring me to the heart of this marvelous landscape.

The Western Rampart

I have now lived on the plains of Colorado for twenty years, but my gaze often focuses on the mountains. The western rampart of the Rocky Mountains dominates the horizon of Fort Collins. Flying west from Denver feels like crossing the roof of the world. Remote, snow-covered peaks mass together, spreading out to the horizons beneath the suddenly small plane.

I gauge the weather and the mood of the day by the look of the mountains. Dense banks of winter clouds often hover above the mountains while the prairie remains sunny, but once the clouds begin to spread eastward we are likely to get snow. Sometimes winter sunsets weave bands of clouds in sumptuous colors through the black outlines of the bare trees, while the peaks lie dark and silent on the horizon. Other days the entire range glows pink at first light.

I escape to the mountains when the pressures of city life become too great, seeking both to calm the frenzy within myself and, as John Muir advised in *Our National Parks,* to "climb the mountains and get their glad tidings." For many years, I followed the pattern established in Arizona, doing summer field research elsewhere and then returning home for the start of the academic year. In September the mountain landscape glows. Aspens explode in bursts of gold, and I walk through golden leaf-falls when leaves swirl across the trail in a gust of wind. The warm, pine-scented woods open to broad khaki-colored meadows of sun-dried grasses. Kinnikinnick berries form claret-colored points on the forest floor, and willows lining the small creeks become undulating lines of burgundy. The shallow lakes of the lower elevations are choked with lily pads. Everything is dry, but full of color. As I climb, water grows more evident. Icy water flows down from unseen lakes into deep, broad valleys creased by avalanches and debris-flow scars. At the base of the peaks lie bowls of small lakes and talus fields closely cupped by wind-blasted krummholz and the sharp, rocky ridges and peaks.

Sunlight and cloud shadows chase rapidly across the steep alpine slopes in the bitter cold wind above the tree line. This is clean, hard-limbed country, all rock and vertical lines but for the tiny patches of green that collect wherever there is a slight fold in the slope. Wildflowers still bloom in these small alpine meadows during September, but the cold, dry winds keep them diminutive. Pale blue and white columbine, yellow stonecrop, blue mountain bell, and red Indian paintbrush rise only slightly above the ground.

I have climbed to these meadows with lungs raw from the cold, thin air, and thighs heavy and aching from exertion, and then looked back to see my companions forming tiny human specks in a vast landscape of rock and ice. But others thrive in this hard world. Muscular bighorn sheep confidently pick their way across the talus. Pikas whistle and cheep at me from their nooks among the boulders. Bold marmots come close, jauntily swinging their cinnamon brown tails as they walk.

Then there is the hike back across the open ridges, with a handful of rugged mountains strewn lavishly along the horizon and an opal moon rising through a sky of lavender silk.

.   .   .

The Rockies are, for me, play and work, school and church, and above all, inspiration. I would like to think that this inspiration comes naturally, and that I would respond to the mountain landscape in the same manner regardless of my time and place. But Rebecca Solnit, in her books *Savage Dreams* and *Wanderlust,* has skillfully traced the history of how Europeans came to admire mountainous landscapes as an outgrowth of the eighteenth-century Romantic movement, which emphasized that the natural world was good and pure and that humans could be inspired by contemplation of primitive landscapes. Now I suspect that at least part of my response to the mountains is conditioned.

European American settlers were so thorough in killing Native Americans in Colorado that today the only reservation in the state is a small area of Ute tribal lands in southern Colorado. Archeological sites at the eastern edge of the mountains suggest that people have lived in the region for at least twelve thousand years. A small quartzite quarry near the timberline in the Rockies has been used for eight thousand years, and rock walls along the tundra were used to drive animals into hunting traps six thousand years ago. Old Man Mountain, a conical granite knob now being crowded by the urban development of Estes Park, may have been a vision-quest site for the past three thousand years. Burial and occupation sites used over the last ten thousand years line the corridor formed by the Poudre River as it flows through the Rockies.

I know a little more about the tribes whose histories have been recorded. Residency on the prairie changed frequently. Comanches moved south from Wyoming toward the end of the seventeenth century to evade the Sioux. Utes lived in the Rockies for at least a thousand years before Europeans arrived, and the Utes joined the Comanches in wars against the Apache and Pueblos, driving the Apache farther south. At the beginning of the nineteenth century, Kiowas moved down from the headwaters of the Yellowstone and Missouri rivers and allied with the Comanches. Kiowas and Comanches were soon displaced southward by Arapahoes and Cheyennes moving southwest from the Great Lakes and upper Mississippi Valley in response to pressure from the Sioux. When the great migrations of miners and settlers began to enter Colorado from the eastern United States, the emigrants encountered Utes in the mountains, Cheyennes and Arapahoes in the west-central Colorado plains between the Arkansas and Platte rivers, Kiowas and Comanches south of the Arkansas, Pawnees along the eastern fringes of the state, and Sioux immediately to the north.

The mountain Utes were the most persistent people. They kept their lands through the centuries of turbulent occupations and invasions of the plains, and they persisted through the European influx of trappers, miners, and settlers. Today they occupy the only Native American reservation in Colorado.

By the time people of European descent reached the Rocky Mountains, the Romantic cult of admiring mountain scenery was well established in Europe. Crossing the plains on their 1805–1807 expedition, the men led by Zebulon Pike gave three cheers at their first sight of the Rockies. "Pathfinder" John Charles

Frémont reserved some of his greatest eloquence for the Rockies, writing in the 1840s with delight at "views of the most romantic beauty" where "Nature had collected all her beauties together in one chosen place." Even the Victorian traveler Isabella Bird, not disposed to regard the plains and new towns such as Fort Collins with much favor, loved the Rockies. She settled in Estes Park for a season, where she enjoyed a little romance with "Mountain Jim" and struggled up Longs Peak.

In 1859, Horace Greeley, after enduring the heat and aridity of the plains, reveled in a summer in the mountains. He wrote of grand, aromatic forests and dancing streams of sweet water. Bayard Taylor, known by his contemporaries as the Great American Traveler, wrote enthusiastically in 1866 of the pure mountain air, crystal water, and alpine scenery. He predicted for his eastern readers that the Colorado Rockies would become an American Switzerland. Katharine Lee Bates, a Wellesley College English professor, wrote "America the Beautiful" during an 1893 visit to the summit of Pike's Peak. And Hamlin Garland, chronicler of the Great Plains, proclaimed, "I am brother to the eagles now!" during his 1890s visit to the Rockies.

Europeans sought to alter the mountain landscape even as they celebrated it. Native Americans probably had relatively little effect on populations of mountain mammals, because Native American population densities were never very high in comparison to those of European Americans. The first European trappers reached the Rockies when members of the Lewis and Clark expedition left the expedition as it returned east to the United States in 1806. Early trappers focused on beaver, mink, and otter. Changes followed swiftly. By the time the Frémont expedition passed through Colorado in 1842–1843, they found mostly abandoned beaver lodges and dams falling into disrepair.

The destruction of beaver colonies throughout the Rocky Mountain West substantially altered the nature of streams in the region. Beaver, sometimes lauded as "nature's engineers," build dams of wood and sediment at closely spaced intervals along streams from tiny creeks to the headwaters of the Colorado River in Rocky Mountain National Park. These low dams can be overtopped and destroyed during the annual high flows of snowmelt, but the beaver quickly rebuild, and the dams store and divert water across the adjacent floodplains during even the highest flows. A river with a healthy population of beaver becomes more diverse and stable. The beaver dams create low steps along the channel in which areas of quiet, ponded water alternate with steeper riffles. These ponds and the overbank flows attenuate floods passing down the stream and reduce the erosion of the stream banks and bed. Fine sediment and organic matter accumulate in the ponded water, providing nutrients and habitat for aquatic insects, fish, and water fowl. As each pond gradually fills with sediment, riverside vegetation encroaches until the pond eventually becomes a meadow and the stream channel shifts its location across the valley bottom. This process of channel migration, combined with the spreading of floodwaters across the valley bottom fostered by the beaver dams, creates a mosaic of bottomland plants and habitat for many species of animals. As trappers reduced beaver populations below sustainable levels, beaver dams fell into disrepair and disappeared, and mountain streams lost the complexity of form and function maintained by the beaver.

The hunters who followed the trappers pursued the bison, elk, wolf, and grizzly bear. Predator control beginning in the late nineteenth century incidentally reduced the populations of smaller mammals such as the fox, skunk, and ferret. The cascade of effects reached the level of the smallest creatures when Old World rats and mice made the journey west with the European immigrants.

Little is known of the effects of European settlement on the small- and medium-size mammals other than the beaver. The list of extinctions for the larger mammals is as harsh as a drum roll. Grizzly bears *(Ursus arctos)* lived in western Colorado and along wooded streams on the plains, remaining in the San Juan and Medicine Bow mountains until the 1920s. They are now extinct in Colorado. Gray wolves *(Canis lupus)* were professionally poisoned in eastern Colorado in the late nineteenth century, and the predator-control agents followed the front of extinction westward to trap, poison, and shoot wolves on the western slope of the Rockies. The gray wolf was extinct in Colorado by about 1940, although at least one wolf has been seen in Colorado during 2007, presumably a wanderer from the reintroduced wolf packs in Montana and Wyoming. The last four wild bison *(Bison bison)* in Colorado were killed in 1897, although a few reintroduced animals now form small herds in the state. The mountain lion *(Felis concolor)* was a bounty animal until 1965, when it became a game animal subject to licensed hunting. The population appears to be stable now, but much reduced from historic levels. The wolverine *(Gulo gulo)* was widespread in the mountains prior to 1890, but now appears to be extinct. The black-footed ferret *(Mustela nigripes)* was wiped out during the war on prairie dogs *(Cynomys* species) but for a small population in Wyoming. Starting in the late 1980s, these animals formed the nucleus for a captive breeding program that now includes a facility in Colorado Springs. Hopefully, ferrets bred in captivity will eventually be reintroduced to the wild in Colorado.

One of the laments in the contemporary destruction of the tropical rain forests is that humans are ignorant of the wealth being lost. So were nineteenth-century Europeans in the American West. In addition to being hunted and enduring increased competition from introduced exotic species, native mammals were stressed by the habitat loss accompanying lumbering, increased forest fires, agriculture, mining, grazing, the building of roads and railroads, and the construction of towns. As I discussed in the book *Virtual Rivers,* the 1859 discoveries of gold at several places in the Colorado Rockies touched off a series of mineral rushes that continued for several decades and gave rise to settlements scattered throughout the mountains. Nearly a hundred thousand people immigrated to Colorado in 1859. This movement of humans was accompanied by deforestation on such a massive scale that whole watersheds were stripped of trees either by cutting or by forest fires started deliberately or carelessly. Burning and tree cutting raged from 1870 until the early 1900s, when forest managers began practicing fire suppression and the first national forest reserves were established.

The ruthless exploitation of Colorado's mineral deposits left a legacy of ruined lands and toxic wastes that continue to poison water and soil. "Gold rush mentality" has become an expression for any selfish undertaking designed to profit the most rapacious individuals at the expense of everyone and everything else. But although the Colorado mineral rushes helped build the personal fortune of millionaires such as Horace Tabor, David May, and Meyer Guggenheim, they also provided opportunities and a sense of community to some of those who came to the mountains. It is easy to vilify the miners when I see mountain streams still poisoned by mining wastes a century later. Anne Ellis's books describing life in the mining towns of the Colorado Rockies from the 1880s through the 1920s put a sympathetic face on the nineteenth-century miners and remind me to attempt to understand the people involved even as I condemn the phenomena they created.

Most nineteenth-century writers praising the beauty of the Colorado Rocky Mountains were not writing about the slopes stripped of trees or the rivers choked with sediment. Some of the writers acknowledged the necessity of bringing progress and civilization to the mountains and of using mountain resources for human gain, however ugly the results. But they did not advise their readers to visit the gold camps for edification; they sent readers to the unmined rocky peaks and the uncut forested valleys.

The writings of Frémont and Taylor and Greeley and Bird influenced me, and they influenced the novelists—Bret Harte, Owen Wister, Louis L'Amour, and Zane Grey—who in turn influenced the writers of screenplays for movies and television shows. Together they created expectations about the western American landscape as a national shrine of nature, freedom, unspoiled beauty, and wilderness.

I also grew up with family photographs and stories. I think of the photograph of myself as a baby happily waving a bottle while loaded into a pack on my father's back as he hiked along some nameless Colorado trail with pines and a lake in the background. Or of the one in which my parents look so slim, youthful, and dashing on a ski slope in the Rockies. Or the one of my mother guiding me down a gentle ski slope at Berthoud Pass when I was barely old enough to walk. My father as a laughing young man with a headful of wavy, dark hair, half-lying in the back of the old station wagon with our German shepherd, stiff pines shadowy in the background. Strings of trout; ground meat frying in a pan with eggs and beans; elk antlers; stories of mountain lions seen near the camp. Camps back in the woods, with only a few

Snowy Mountains, Wyoming

other people and much laughter. The Medicine Bow Forest, the Snowy Mountains, the Never-Summer Range: these names took on a mythical quality as I grew up hearing tales of adventures there. Memories of the stories accompanying photographs taken before I was born are more vivid and deeply ingrained in me than memories of my own adult visits to these places. I grew up with great expectations.

When I moved to Colorado, the mountains met those expectations. I detoured around the cities and the sites scarred by mining. The forests cut during the late 1800s have now grown back. They are mostly even-aged stands of trees sixty to a hundred years old, with less diversity of age and species than the forests that existed prior to the arrival of Europeans. But they are beautiful.

.   .   .

TO REORDER YOUR UPS DIRECT THERMAL LABELS:

1. Access our supply ordering web site at **UPS.COM**® or contact UPS at 800-877-8652.

2. Please refer to label #02774006 when ordering.

02774006 RRDR

Drop-Off Package Receipt: 1 of 1

DROP-OFF LOCATION:
The UPS Store #3898
20126 BALLINGER WAY NE
SHORELINE WA 98155-1290

DROP-OFF DATE/TIME:
Mon 24 Jul 2023 11:40 AM

ESTIMATED PICKUP DATE:
UPS   Mon 24 Jul 2023 1p|

TOTAL PACKAGES: 1p|

TRACKING/REFERENCE	CARRIER & SERVICE	WEI
1Z74Y86F9092159314	UPS Ground	31 lb 12.0 oz

By late autumn the early morning sunlight shines through the frail silver candelabra of frosted aspen tips. Across the valley, each conifer stands out individually, highlighted by a coat of snow. Diamond dust fills the air as the crystalline snow filters down through the sunbeams. The hollow, staccato raps of a woodpecker and the trilling of a squirrel echo over softly imprinted deer tracks. Ponderosa trunks glow ocher against the snow as the sun rises. Snow bombs drop from the trees as the day progresses, cold down the back of my neck. The ground snow springs to life as a multihued fire. Meadows glitter with smooth-swept flakes of diamonds. Puffed snow collects on the branches, crystalline snow round the trunks; icicles bulb and stretch from the needle ends. Thistles up to their necks in snow nod beneath domed white caps. A bald eagle hunches against the rising wind, his head seemingly snow frosted. At twilight the aspens resemble gray winter ghosts with dark, snow-crusted eyes. They form gray paths down from the sky by which the clouds come to the snow, deep and silent. Snowflakes slip down between the planes of air.

When I pause while skiing, the forest is so quiet I can hear the soft crinkling sound of snow falling on my jacket. The footsteps of the dogs accompanying me vibrate and echo on the solid base of snow beneath the powder. Trees shifting in the slight wind make a rhythmic *c-r-e-a-k, c-r-e-a-k*. I look up and a snowflake lands on my nose in a sudden spot of cold, then melts into wet warmth. Soft calls of juncos and chickadees are muted by the snow thickening the air, and even the trill of a squirrel scolding the dogs doesn't carry far. Through eyelashes sticky with snow, I appreciate the colors of winter: textured white snow, somber green branches of pine and fir, warm reddish bark of ponderosas, blue-gray sky the hue of ice on a deep pond. In summer the rivers and creeks here are the color of herbal tea. In winter the creeks look black, and the rivers flow pale jade-green over the ice, and forest-green in the deeps.

Then come February mornings of high, thin clouds and dry winds that blast over the ridges and down the valleys, nearly taking me with them. The mountains form a winter sere landscape of wind-twisted cedar and grizzled dry grasses, and the ground beneath the trees is thinly scabbed with snow. Other February days begin in bluebird mornings of white snow glowing beneath blue sky, and wind-blown plumes of snow and cloud flagging silently off the crests of the peaks. The slopes are pocked with rabbit and fox tracks. Clark's nutcrackers and mountain chickadees call from the frosted pines. The aspens are a warm cream color. They cast latticed shadows on the snow and form a delicate screen for the solid peaks beyond. Skiing through aspens I sometimes laugh aloud in sheer joy at being able to live in such a place. It is like a fairy tale come true.

· · ·

Other people seek to live the same fairy tale. Because mountain scenery continues to inspire people, mountain communities continue to grow. Some of the communities are mining towns transformed into ski resorts; others are vacation or full-time retreats for those weary of traffic, air pollution, and crime. The continuing influx of people, with their roads, utility lines, water and sewage requirements, and noise, further changes mountain landscapes and creates hazards for those who would be brothers to the eagles. I got a closer look at these hazards when my student Mario Mejia-Navarro studied Glenwood Springs. Although I jokingly refer to Glenwood Springs as a poster child for environmental hazards, the city is unfortunately not unique in that respect, but represents many of the rapidly expanding mountain communities.

Glenwood Springs lies within a narrow valley at the junction of the Roaring Fork and Colorado rivers. It seems like a precarious place to live. Floods along either or both rivers may inundate the low-lying portions of the town, as they did in 1918, 1938, 1957, and 1963. Debris flows and landslides from the steep slopes that closely hem in the town may bring thousands of tons of mud and boulders cascading onto the outer edges of the valley, as they did in 1903, 1917, 1929, 1936, 1937, 1943, 1963, 1977, and 1981. The town is built on gently sloping fans created by numerous debris-flow deposits over tens of thousands of years. Every so often this poorly consolidated sediment settles, cracking the foundations of buildings and buckling the sidewalks. And then of course there are the periodic avalanches and forest fires, most recently the severe fires of 2002, which further destabilize the hillslopes.

The most recent major debris flow occurred in Glenwood Springs in 1977. This flow spread across two hundred acres of the city. Yet so short is human memory, and so poor is the average citizen's understanding of landscape processes (or perhaps so reckless is their attitude toward probabilities), that many people in Glenwood Springs ignore the potential for further damage from debris flows. Mario was commissioned by the county and city planners of the region to develop a model of geologic hazards in the city. At present, the city has zoning recommendations, but these are not enforced. Houses backing up to the valley walls are supposed to have specially reinforced foundations and no doors or windows at the back of the house. Mario and I walked along the steep valley side slopes, crossing dry stream channels choked with loose debris that could be mobilized into a debris flow. We looked down onto beautiful houses with large expanses of glass at their backs. I felt as if I were watching a loaded cannon pointed at those houses.

We walked city streets laid out in a traditional American grid pattern. One part of the grid runs parallel to the Roaring Fork River, and the other part runs at right angles to the river and the slope of the valley walls. These latter streets dip steeply from the valley sides down to the river and form natural conduits for water and sediment coming off the valley walls during a debris flow. During the 1977 debris flow, some of the streets actually channelized sediment. Mario and I visited the Little People's Place, a day-care center with a charming entryway sized for children, through which adults would have to stoop low or even crawl. This entryway pointed upslope and would certainly be choked with sediment during a debris flow, potentially trapping anyone who was inside.

Those who admire mountain scenery and wish to live in it must realize that the scenery is not static. The mountains and valleys were not created in an instant and then left unchanging, nor does change occur at such immensely long timescales that an individual human will never observe it. Large changes in the landscape are compounded of successive small changes that can be immense for a human. In the first few months that I lived in Colorado, I read a newspaper story about a couple who had just finished building their retirement cabin at the base of a picturesque rocky hillside in the mountains. A rock fell from the hillside during the couple's first night in their cabin, crushing and killing both husband and wife. Swiss peasants living in the Alps knew that glaciers periodically moved, and they refrained from building houses at the bases of glaciers despite the spectacular views. Those who live in the "American Switzerland" need to be similarly aware.

Awareness may not come readily. It implies limitations where Americans have traditionally been unwilling to acknowledge them: in the choice of site for a home or a community, and in how commonly held resources are used by individuals or small groups of people. But awareness of limitations is growing slowly. What European settlers might have perceived as unlimited abundance two centuries ago is now clearly limited. Low-density housing and the network of unpaved roads necessary to sustain it carve up many hillsides and valley bottoms that were unsettled just a decade ago. Economically successful mountain communities grow to capacity as land and housing prices rise, then spawn satellite bedroom communities for service workers unable to afford living costs in the primary community. Parking areas at trailheads spill cars onto the shoulders of adjacent roads, and campgrounds fill early in the day.

Mining and logging are now limited in the Rocky Mountains of Colorado, partly for economic reasons, partly because society no longer considers such activities acceptable in the midst of ski resorts and expensive homes. But I think fundamental attitudes toward the natural world have not really changed. Now the people of Colorado mine the mountain scenery, parceling it out and selling it off, inevitably reducing the very qualities that people admire and seek out. The ability to journey beyond sight, sound, and other traces of other humans is now very limited—if not gone altogether—in the Rocky Mountains of Colorado. Interstate highways and secondary roads form a dense network across and through the mountains. Many of the routes follow trails established by migratory animals and then Native Americans and

European "explorers." Many of these roads also follow the rivers, however deep and narrow the canyon that the river has carved.

. . .

I, too, follow the deep, hidden paths where water flows down and tourists flow up—skiers in winter, hikers and anglers in summer. Following the windings of the water paths in winter, I find ice. Clear, sharp days of ice beneath winter-bare trees with thin shadows. Ice milky white and brittle as spun glass. Turgid green ice, thickened where the black eddies pool beneath knotted tree roots. Congealed ice foam in white and pale aquamarine. Ice frozen into rapids and eroded into icicle-encrusted ledges. Sleek ice terraces shiny as Jello. Ice furrowed and molded when water froze in the act of rippling. Ice shot through with bubbles. Ice pressured and ridged. Stegosaurus plates of ice bristling from a branch. Brown leaves frozen deep in the ice, set and preserved like museum specimens. Ice bells hanging down from a ledge into the water. Ice rainbows of winter. The joys of life in a cold climate.

So much depends on ice. Water that has slipped through the hydrologic cycle across half a world briefly comes to rest, expanding as it freezes, so that it floats. This the fish count on, winter after dark winter, deep and still beneath the lid of ice. Ice embodies contrasts: fragile as old bones when it is thin and bubbled over a puddle, hard as a rock when a skater falls on the thick, deep-booming ice of a pond in midwinter. The pressure of a small human body focused on the slim blade of a skate is sufficient to melt ice. Yet when it flows in great glacial masses, ice drags boulders that carve out whole landscapes, scraping down valleys and pushing up moraines. A single hairline fracture through a rock into which water can seep and freeze, expanding as it chills, is

A delicate, layered ice shelf projecting above the river's edge

all that ice requires. With time, the most solid granite will shatter from the ice's growth. The high country of the Rockies lies rubbled from the work of ice. Roads coated with early-morning ice and shadows are by late afternoon littered with chunks of rock dislodged from the valley walls by melting ice.

With the coming of spring, the snows grow heavier with moisture and the rocks fall frequently. Spring snow comes in conglomerations of heavy flakes. Branches laden with snow resemble nodding, white-furred creatures as they bob and sway in the sudden release of their loads. The ridge crests, the breeding ground of avalanches, are lost somewhere in the clouds above. The timberline is a whitescape where the blowing snow grizzles the slim, wiry spruces. The sun forms only a small patch of faint white light beyond a thousand layered tissues of snow and cloud. There I can be very much alone in a silent world of shifting white, lungs and legs working like bellows, mind stopping short in silent contemplation.

By April, meltwater flows in every depression and crevice as the mountains slough off the long winter. I slog up the trails through the wet, heavy snow of spring, past aspen groves just ready to bud.

.   .   .

The annual cycle of snow falling and then melting is crucial not only to the mountains of Colorado but also to the remainder of the state and to most of the neighboring states. People living in the Rocky Mountain West absolutely must have water from the mountains. But too much water creates floods that destroy structures and human lives. Too little water creates a regional panic that inspires extreme measures. After four years of drought at the start of the twentieth century, local and state officials in Colorado resorted to cloud seeding to increase snowfall, removal of beaver and beaver dams to facilitate the downstream flow of water, and local thinning of forest stands to increase runoff from hillslopes. The governor proposed the "Big Straw" project to pump water back upstream from the border of western Colorado so that Coloradoans, rather than Californians, could use the water. Current proposals involve piping water from the Green River in Wyoming to the Colorado Front Range urban corridor and building new reservoirs to store water in the South Platte River basin. Every effort is made to alter the supply of water, and relatively little to alter human use of water. In Colorado we do everything but change our own behavior when drought threatens water supplies.

For Coloradoans, the mountains serve as a magnetic center that deflects our diverse lines of thought. I go repeatedly to this magnetic center, this western rampart that defends me against drought of the land and drought of the soul. I come to this region from a hundred different points, piecing the fragments into a coherent whole as I hike and ski in all weathers and seasons. In this richly textured landscape, peaks and valleys may be only a few miles from each other yet look completely different. I savor names that invoke imagery and history: Box Canyon and Thunder Pass on the Never-Summer Trail; Ruby Jewel Lake in the Medicine Bow Range; Cirque Lake; Timberline Lake in the Comanche Peak Wilderness; Nokhu Crags; Glacier Gorge; Sky Pond; Mummy Pass; Roaring River; Ptarmigan Mountain; Ouzel Falls. Gradually I am fitting myself into the groove of these mountains and coming to understand their times and places. I hope someday to know this country.

Where the Winds Live

The windows of the house are open to admit the beguilingly soft, warm air of late spring. Along the rivers, the cottonwood trees are starting to lose their seeds in blizzards of down that leave a fluffy white froth everywhere. Out on the prairie, the slopes flash colors that change every few days as new wildflowers come into bloom. White sand lilies and yellow wild parsley give way to blue flax and lavender lupine. Yucca send up slender stalks crowned with creamy white blossoms. The fluid call of a meadowlark coming in on the breeze reminds me that I live on the plains. I do not always remember this basic fact. When I travel, I tell people that I live in Colorado. They immediately think of the Rockies.

The Great Plains were the threshold during my childhood trips to the West. The plains themselves were not particularly exciting. All I really saw from the interstate was mile after mile of crops. I liked the feeling of spaciousness, but the landscape seemed monotonous. An occasional glance out the window of the car sufficed. My interest in the prairie stemmed mainly from the fact that

Sand lilies, Colorado plains

it represented the start of the West, where the Indians had hunted bison and the cowboys had trailed cattle herds. Somewhere in my crossing of the plains I would see the mountains, and that was what I waited for.

My sight remained fixed on the mountains when I moved to Colorado. The plains were the great expanse to the east that provided a buffer against the more populous part of the United States. They were the transition zone across which rainfall steadily decreased from the abundance that nourished deciduous forests, to amounts that supported tallgrass prairie, to the pittance that grew bunchgrasses and cacti in eastern Colorado. The plains seemed more of a historical setting than a contemporary land-scape, plowed as they were from interstate to interstate. When I had a day to spare and wanted to hike or ski or camp, I went west to the mountains.

I am not the first to have underestimated the plains. Many people came here looking for something else. I am convinced, after spending time in Australia, New Zealand, and the United States—regions recently colonized by Europeans—that people have a basic penchant for importing the known without stopping to consider the details of a new environment. How else to explain the European attempts to graft imported plants and animals and human cultures onto new lands? So many of the attempts backfired in unforeseen ways: imported rabbits breeding to plague proportions in Australia; English hedgehogs destroying the habitat of ground-nesting birds in New Zealand. Honeybees, earthworms, sparrows, starlings, brown trout, tamarisk, mullein, cheatgrass, leafy spurge, zebra mussels, and hundreds of other species deliberately or accidentally introduced to North America have wreaked havoc on native ecosystems. Why? So that poor people could eat rabbits in Australia. So that English settlers in the Antipodes could enjoy the sight of an animal familiar from home. So that European colonists in the New World might harvest honey, hear familiar birdsong, and so forth. At a level much more fundamental than the introduction of an individual species, the European colonists sought to import ways of life developed in Europe, ranging from residing in one spot for long periods of time rather than living as nomads, to raising crops that required much more rainfall and consistent moisture levels than the newer, drier lands provided.

These ways of life were not well suited to dry grasslands. Most organisms inhabiting grasslands, from insects to birds to grazing ungulates such as bison, to indigenous humans, are nomadic to some degree if given sufficient freedom. Migration probably developed in response to the limited moisture and available nutrients in the dry grasslands, which can provide a very rich environment if you keep moving. Dry grassland soils form an enormous reservoir of carbon and nitrogen. This reservoir has accumulated over tens of thousands of years, and only a small portion of it actively moves through the tissues of living organisms at any given time. Anything that interrupts the tight cycling of nutrients, such as removal of the sparse vegetation or erosion of the topsoil, releases carbon and nitrogen and drains the reservoir.

The first Europeans to reach the grasslands of the western United States did not recognize these limitations. Spaniards entering southeastern Arizona from Mexico were entranced at grasses that grew as high as a horse's back. The Spaniards and their Mexican successors promptly stocked the grasslands with densities of cattle higher than the native bison had ever achieved. The grasslands could not sustain this intense grazing. Today the range remains much poorer than it was at the time of first Spanish contact.

European Americans coming west made similar mistakes two hundred years later. John Charles Frémont's eloquent 1845 report of his exploring trips describes abundance, not limitations. Frémont wrote of elk, deer, antelope, "troops of wolves" following the bison, prairie dog villages "so thickly settled that there were three or four holes in every twenty yards square" along two miles of river bottom, and even grizzly bears. But it was the bison that dazzled him. He described them "swarming in immense numbers over the plains, where they had left scarcely a blade of grass standing." The massed bison resembled large groves of timber at a distance. Herds of more than ten thousand animals covered the plains as far as Frémont could see. He watched bison hunts, describing what "seemed more like a picture than a scene in real life." Within four decades the seemingly limitless bison herds were hunted nearly to extinction.

The great era of the open range depicted in so many movies lasted from about 1867 to the 1880s, when barbed wire and homesteading restricted the movement of cattle herds. During that brief period, huge cattle drives followed the Chisholm Trail from Texas to Abilene or Dodge City. The cattle industry boomed after the national financial panic of 1873, and the resultant overstocking led to a crash in 1885–1886, as thousands of cows died during a hard winter. The remaining ranchers invested in fences and windmills, a process that historian Walter Prescott Webb described as converting ranching "from an adventure into a business."

Sodbusters displacing the cowboys saw in the native grasses a promise that the land would support wheat and barley and rye. Settlers flocked to the Great Plains after the Civil War in response to encouragement by boosters such as editor Horace Greeley, who uttered the famous admonition "Go west, young man!" Earlier western migration had aimed at California and the Oregon Territory and avoided the interior grasslands that explorers Zebulon Pike and Stephen Long had called the Great American Desert. Now would-be farmers turned to the grasslands, spurred on by a series of federal acts that represented a grudging acknowledgement of the region's dryness. The 1862 federal Homestead Law allowed families to take up 160 acres, but it was not until the sale of the first piece of Joseph Glidden's barbed wire, patented in 1874, that settling the plains really became feasible for the homesteader. Cheap, effective fencing that staked out free homesteads revolutionized land values. The

Timber Culture Act of 1873 tried to extend the Midwestern woodlands to the West by giving land to those who would plant it in timber. The 1877 Desert Lands Act enlarged allotments in arid lands, from 160 to 640 acres. The 1909 Enlarged Homestead Act granted 320 acres in semiarid regions.

These land allotments remained woefully inadequate for any form of agriculture in the drylands, where John Wesley Powell estimated in his *Report on the Lands of the Arid Region* that a minimum of 2,540 acres was necessary for a family farm. Bernard De Voto chronicled in outraged detail the massive land frauds that resulted as wealthy livestock owners and corporations used the letter rather than the spirit of the laws to gain title to extensive acreages.

The sodbusters imported varieties of grain that needed much more moisture than provided by the natural rainfall. They then had to divert water from stream channels or pump it from beneath the ground to keep the crops growing. People suffered terribly during the droughts that periodically settled over the Great Plains for years at a time. Each period of drought caused farming families who had put every ounce of energy into settling their land to abandon their homesteads. Ole Rolvaag, Willa Cather, Mari Sandoz, and others wrote eloquently of the emotional and physical toll of establishing a farm in the dry plains. Yet each period of renewed rainfall following a drought saw hopeful new settlers taking up abandoned farmlands once more. Wallace Stegner's famous description of the West, in his *Wilderness Letter,* as "the geography of hope" encompasses layers of irony.

The federal government established the national grasslands following the great droughts of the Dust Bowl partly in an effort to keep people from farming these agriculturally marginal lands.

But when World War II raised wheat prices, farmers descended on state legislatures demanding the repeal of laws giving soil conservation districts regulatory powers over the agricultural practices of their members. In an essay collected in *De Voto's West,* De Voto quoted a Soil Conservation Service employee in Washington, D.C., as saying in 1947, "I can't tell you when the next dust bowl will come, but I can tell you where it will come." The man pointed to a Colorado county on a wall map. A subsequent drought bore out his predictions, and in 1953 President Eisenhower declared the county a disaster area eligible for relief. As De Voto wrote, "We would bail out the sacred principle of land destruction once more."

The farmers who stayed on the prairie gradually consolidated smaller holdings into larger and larger ones that required massive investments of capital in machinery and annual purchases of seed, fertilizer, and pesticides. Farmers applied ever-greater amounts of fertilizer as applications of water leached stored nutrients from the soil, in effect artificially maintaining the crop fields at the expense of everything else around them. The excess fertilizers flowed into streams along with runoff and seeped into the underground waters. Nitrogen and phosphate caused algal blooms and depleted oxygen levels in the streams, killing off aquatic organisms. Water draining from farm fields ultimately flushed so much nitrogen into the Mississippi River system that the Gulf of Mexico began to experience huge dead zones, where the water was so depleted of oxygen that nothing could survive. Nitrogen in the wells of farming communities caused lymphatic cancer and, by preventing blood from carrying sufficient oxygen to the body's tissues and cells, blue-baby syndrome.

Agricultural monocultures in the form of vast fields devoted to one variety of a crop were beset by insects and weeds that farmers wanted to be rid of as cheaply as possible. Increasingly intense applications of pesticides contaminated local soil, water, and air with toxic compounds that decayed only over a period of decades, and were eventually shown to cause cancer and a host of other health effects in various organisms, including fish, birds, and humans.

None of these consequences were foreseen or intended. On the contrary, the pioneers worked incredibly hard to create a better life for themselves and their descendants. They saw themselves as the vanguard of civilization, reclaiming wastelands for a strong and growing United States. Unfortunately, they never really saw the environment that they immediately set out to change. They did not recognize how native organisms and indigenous peoples had evolved in response to the climate and soils present on the Great Plains.

Such recognition would have saved much subsequent heartache, but, as Wallace Stegner wrote in *Beyond the Hundredth Meridian* when discussing the booster mentality, "to the Gilpin mind facts are not essential. . . . What is more essential is vision." William Gilpin was the first governor of the new Colorado Territory established in 1861. He had been a soldier and explorer with Frémont in 1843. By the 1860s he had settled down to being a newspaper editor and used his paper to preach about the attractions of a new land with unlimited opportunities. Gilpin envisioned a West more populous than the eastern United States. Where aridity seemed to contradict him, he overrode the fact with vision and ignored the fundamental limitations of sparse water supplies.

Part of the Gilpin attitude stems from an ingrained Judeo-Christian abhorrence of desert, which is only worthwhile if it can be made to blossom as the rose, a claim made by the Mormons for the Utah desert in 1849 and subsequently picked up by other Westerners. But, as De Voto tartly exclaimed in an essay collected in *De Voto's West,* "the West has not made the desert blossom. By means of the most formidable engineering works man has ever constructed, it has transferred portions of the mountain snow pack to minute areas that lie along the edge of the desert."

Part of Gilpin's attitude stemmed from the fact that boosters such as he never wanted to admit that any portion of the grassland was arid or semiarid and thus less than agriculturally optimal. Instead the boosters claimed that "rain follows the plow." The sodbusters taking up land with such hopes did change the regional ecology, if not the climate. As farmers plowed up the native grasses and suppressed fires, the prairies gave way to crops, introduced pasture, and small woodlots in some areas, all supported by mining the underground water supplies or diverting surface streams.

These changes occurred with astonishing rapidity. The heyday of the Old West, the era of mountain men, cowboys, and gold rushes that gave rise to such a wealth of stories, lasted less than half a century. Mountain men pursued their rugged lives between about 1825 and 1840. The Pony Express delivered mail from April 1860 to October 1861, when the cross-country telegraph was completed. Mining rushes sparked by the 1849 discovery of gold at Sutter's Mill in California were largely over by 1890. The immense bison herds were hunted to extinction in twenty years. The cattle drives lasted twenty years. Almost immediately the West assumed an epic quality, and individuals who had been young at its opening found themselves ghosts of the past before their own deaths. In *A Daughter of the Middle Border,* Hamlin

Garland describes his father's wonder at a lifetime that included work as a young laborer grading the first railway in Maine, and travel as an old man across the "Great American Desert" of the western plains in a Pullman railroad car that served steak.

Some have considered the speed and thoroughness of change on the Great Plains to be a sign of progress and the inherent justness of American civilization. Deborah and Frank Popper, however, proposed in the 1980s that the declining human populations of parts of the Great Plains be removed in order to establish a Buffalo Commons devoted to restoring pre-European ecosystems. Their suggestion was met with howls of indignation from residents of the plains who saw it as devaluating and forgetting their pioneer ancestry.

I too question whether the "civilization" we have grafted onto the plains indicates progress. My perception of the plains changed dramatically when I moved to a house adjacent to a 650-acre shortgrass prairie preserved as open space by the city of Fort Collins. Exploring the prairie in different lights and seasons led to my curiosity about what I was (and was not) seeing. Background reading inspired me to further scrutinize the prairie, and this led to a revelation. What can appear monotonous and boring when compared to the obvious drama of the mountains holds, on closer inspection, a wealth of fascinating detail.

Bunchgrasses adapted to aridity and limited nutrients form centers of life among the surrounding patches of exposed soil in the shortgrass prairie. Although the grasses grow only a few inches above the ground, they can send roots down several feet. Soil fungi colonize the root network. Plant litter shed by the grasses feeds other microscopic organisms. A fraction of an ounce of prairie soil in the vicinity of the grass roots may be teeming with more than a billion bacteria, ten to twenty million fungi, three million algae, a million protozoa, and fifty nematodes. A few feet away, the limited input of nutrients keeps the population density of soil microbes much lower. Habitat and nutrients that the grasses provide also support more visible organisms, such as earthworms, mice, prairie dogs, rabbits, antelope, and bison. Rodents and rabbits provide food for snakes, raptors, and coyotes. Each creature has evolved its own strategies for accessing and thriving on the prairie's limited resources. The colonies of prairie dogs historically reached 150 miles wide and 250 miles long, with an estimated four hundred million individuals in one colony. Frémont and other early European American observers wrote of bison herds that took hours to cross the observer's field of view. Life this abundant can be maintained only because each organism gives back to the prairie. Each creature's waste, and eventually its body, provides nutrients that other organisms recycle into a tightly held pool of carbon and nitrogen. Whereas other ecosystems, such as tropical rain forests, hold the majority of nutrients within the tissues of living organisms, the shortgrass prairie community has gradually built up a reservoir of nutrients in the soil.

Human patterns of resource use disrupt this recycling system by extracting far more nutrients than are returned to the prairie. People of European descent in particular have simultaneously underestimated the biological wealth of the prairie ecosystem and overestimated their ability to withdraw resources from it. As a result the prairie, with its massive habitat and species losses, is now one of the most endangered ecosystems in North America.

I did not know of the prairie's fascinating details or historical changes until I made an effort to read about the prairie and observe the landscape carefully. When I first moved to

Black-tailed prairie dog at the entrance to its burrow on
the shortgrass prairie of eastern Colorado

Colorado from Arizona, the plains and the mountains seemed a
little overvegetated, a little too green and busy. When I returned
to Colorado one summer after six months in Japan, the landscape
seemed blasted and barren, too dry to support a continuous, thick
cover of plants. Several years later I returned after living in New
Zealand for three months. I arrived home late in March, just after
a heavy snowstorm had spread a lovely white mantle across the
mountains. Driving home, I admired the mountains to the west,
but I was equally excited to see the first stirrings of new growth
on the brown prairie. I embraced the partnership of mountains
and plains and felt delighted to return to both of them.

When I first began to live on the Colorado prairie, the constant
wind made me uneasy. In Ohio, strong wind usually presaged
something unpleasant: a tornado or winter blizzard. It took me
years to accept that strong winds are always present on the prairie,

whether they are summer winds blowing warm, moist air across the plains from the Gulf of Mexico, or winter chinooks sweeping warm, dry air down across the mountains. The strong winds presage change, but change is also constant and not necessarily unpleasant. I have finally changed my perceptions and acknowledged that, like the winds, I live on the plains.

PART 2

Inheriting the Past

Colorado Burning

Fire happens. Fallen branches, dropped needles and pine cones, and partially decayed leaves and plant stems accumulate on the forest floor. The dry heat of summer desiccates the fallen material. A lightning strike ignites a dead snag, a passing motorist tosses a lit cigarette, or a camper lets the wind carry the still-glowing embers of a campfire into the dry duff beyond the fire ring. Something ignites. Flames spread to the surrounding trees and grasses.

Once the fire starts, it becomes unpredictable. The flames may creep along the ground, consuming the dry material on the forest floor but never rising high enough to seriously damage living trees or green shrubs. A narrow creek or even a foot trail may provide a break sufficient to stop or turn the fire. Or the fire may become an inferno, creating its own climatic conditions as it tears through the landscape in erratic streaks, roaring several hundred feet up a slope in a few seconds, jumping rivers and roads, leaving a charred swath half a mile long in which everything is dead, with unburned forest just a hundred feet away. At their peak, these crown fires consume mature trees like Roman candles erupting.

The nature of any fire depends on the weather conditions and the supply of combustible material, or fuel load. Hot, dry, windy weather promotes more intense, dramatic fires. Ground fires can burn slowly for years under cooler, moister conditions. Fuel loading also depends on weather, because precipitation and humidity levels help govern the amount of moisture stored in live plants and dead plant litter. The time that has elapsed since a forest last burned or was somehow cleared of dead organic matter, and the types of plants growing in the forest, also govern fuel loading.

I never thought much about any of these things when I lived in Ohio. Fire was something we started in a grill to roast hot dogs, or in our big flagstone fireplace on Friday and Saturday nights. The woods in Ohio just didn't burn. First of all, there weren't any extensive tracts of forests left to burn, and then the weather was usually so humid that the forest was too damp to burn much.

My perception of fire changed as soon as I moved to the western United States where, one way or another, the forests burn regularly. Sometimes the fires are deliberately set for controlled burns designed to consume accumulated fuel before it reaches a dangerous level, or to create conditions favoring plants that have evolved with periodic fire. More often, the fires result from lightning strikes or unintentional human acts. The existence of unplanned fires is easily one of the most inflammatory issues in the West.

.   .   .

I first witnessed the effects of wildfire while studying debris flows in the Huachuca Mountains of southern Arizona. The steep slopes of the Huachucas rise from the desert lowlands. Their grassy oak chaparral gives way to stunted woodlands of pinyon pine and juniper and to tall ponderosa pine forests at the higher elevations. Like other mountain ranges in the desert, the Huachucas form an oasis of green shade and relative coolness, and small communities cluster around their base. In June of 1988 a forest fire crossed the adjacent border from Mexico and burned up the southern slope of the Huachucas and across the crest, descending the northern slope toward a group of houses tucked along the base of the range. The fire was so intense that it completely destroyed most of the vegetation in its path, leaving charred trunks above a soil stripped of ground cover. The mountain range is largely Forest Service land, and flame retardant dropped from helicopters stopped the fire before it reached the houses. But the homeowners had further natural disasters in store.

A sufficiently hot fire vaporizes chemical compounds from plants and drives them a few inches down into the soil, where they precipitate in a waxy layer that prevents rainwater from penetrating. About a month after the Huachuca fire, a summer thunderstorm poured almost an inch of rain onto the denuded slopes. The water filtered down to the waxy layer and saturated the overlying soil, which began to flow downslope. The moving soil concentrated in the dry stream channels, where it gathered sufficient momentum to rush down the channels as a debris flow, a slurry of mud and water that can reach speeds of up to thirteen feet a second and carry house-sized boulders. The people living at the base of the mountains said that the rain lasted only fifteen minutes that day, followed by the noise of a freight train coming down the mountain. No one was hurt by the debris flow, but it caused minor damage to some houses and filled backyards

Phil Pearthree in an erosional reach of the debris flow in the
Huachuca Mountains, Arizona

and corrals with mud and boulders. The homeowners blamed
the Forest Service for the debris flow, because the Service had
controlled but not extinguished the fire. The Arizona Geological
Survey was called in to investigate, and my friend Phil Pearthree
recruited me to help.

From September to April we walked slopes where the fire had
left standing pieces of charcoal, comparing them to the unburned
slopes with tall pine groves and thick layers of duff underfoot.
One day a thunderstorm caught us on the burned slopes, and we
watched as pelting rain swiftly eroded the loose, exposed sediment into small, parallel channels. Each rainstorm flushed a little
more sediment from the slopes, and the most intense rainstorms
created a fire-hose effect that concentrated water in new channels
and stripped overlying sediment down to bedrock. From miles

Phil Pearthree along a depositional reach of the debris flow in the
Huachuca Mountains, Arizona

away, these skeletal white channels stood out among the darker
browns and greens of the surrounding sediment and unburned
areas. Over a period of weeks, we surveyed these new channels
down which the debris flow had rushed, measuring the loca-
tions and amounts of sediment carried away and deposited by
the flow. Following the channels downstream into flatter reaches,
we scrambled over the shattered trees and huge piles of boulders
where the debris flow had begun to slow down and deposit sediment. Here we found eroded stream banks where the 1988 event
had exposed older debris-flow sediments, evidence that such flows
have occurred in these mountains for tens of thousands of years.
The houses affected by the 1988 flow were built at the base of a
gentle slope composed primarily of older debris flow deposits.

As the study unfolded, Phil and I developed a conceptual

model of the response of hillslopes and stream channels in the Huachucas to fires and debris flows. Following a debris flow, vegetation gradually regrows. Sediment from the hillslopes slowly accumulates in the dry stream channels, forming swale-shaped troughs stabilized by grasses and shrubs. Leaf litter accumulates on the forest floor until, one dry summer day, something starts a fire. If these fires occur frequently—perhaps once every ten years—they will probably not kill the trees. They will destroy the undergrowth, exposing the hillslope enough to allow some sediment to be washed into the stream channels during rainstorms and causing small debris flows. In 1988 we observed areas that had also burned in 1977, and these did not have large debris flows. On the other hand, if a long time passes between fires, and sufficient fuel accumulates on the forest floor, the next fire may be so intense that it kills all the vegetation and creates the water-repellent soil layer that fosters large debris flows.

The Huachuca slopes that burned in 1988 were ready for a fire. If the Forest Service had suppressed that fire, the stage would have been set for a later, more disastrous event. What the homeowners at the base of the mountains saw as a single, preventable catastrophe struck me very differently. I viewed it as the latest occurrence in a long and continuing sequence of landscape evolution in the Huachuca Mountains, and an occurrence that was fairly predictable given some knowledge of the range's recent geomorphic history.

In my subsequent work with graduate students elsewhere in Arizona and in Colorado, fires continued to stand out as a primary driver of landslides and debris flows on hillslopes. After a severe wildfire burned much of the Buffalo Creek drainage southwest of Denver in 1995, my student Casey Clapsaddle investigated the response of small streams that were abruptly overwhelmed with tons of sand and gravel from the newly denuded hillslopes. Casey and I fastened waterproof boxes containing cameras on trees next to the creeks. Internal timers triggered the cameras to take photos every twenty-four hours. Viewing those photos, we saw waves of sediment up to ten feet thick deposited along a tiny creek as a summer thunderstorm created a flash flood. Sometimes the sediment was completely removed a few days later as another flash flood scoured the channel. All of this sediment moved downstream to concentrate in the larger channels, where it contributed to severe flooding that destroyed bridges and homes, as well as filling one of Denver's water-supply reservoirs with sand and decaying logs.

Very intense rainfalls can disassemble hillslopes into debris flows and landslides, as happened during the July 1976 flood along the Big Thompson River. But the most massive and widespread landslides usually come with the first rains after an intense wildfire has killed hillslope vegetation.

. . .

Several things can happen during a very hot fire. Intense heat vaporizes low-growing shrubs, grasses, and forbs, leaving little evidence of their presence. Trees are killed, although they may remain standing for years, their roots slowly decaying underground. Sometimes a fire follows the tree trunk underground, burning out the roots and leaving hollowed pipes in the soil. Boulders and bedrock outcrops do not efficiently conduct fire's intense heat, so surfaces heat up while interiors remain cool. This temperature differential cracks the rocks, causing slivers and chunks to spall off the surface.

Human metaphors about their strength and permanence notwithstanding, rocks gradually crumble into sediment. Organic acids excreted by plants as diverse as trees and algae weaken the rocks. Expansion and contraction from countless cycles of heating and cooling, freezing and thawing, and wetting and drying, all break up the rock as steadily as jackhammers demolish concrete. Over time a layer of loose sediment settles on a slope.

This sediment is effectively anchored by the roots of plants. The portions of the plants growing above the ground break the impact of falling raindrops, and the layer of plant duff on the ground acts like a sponge, allowing rainfall and snowmelt to gradually filter down into the soil. When a wildfire kills the vegetation on a hillslope, suddenly all of the sediment loses its mooring. If the exposure of bare soil reaches or exceeds about 30 percent of the ground surface, sediment movement increases substantially. Gravity drags sediment downhill, lubricated by any recent rainfall or snowmelt. Water running off the surface creates millions of tiny channels called rills. In many burned areas, flow in these rills moves much of the sediment carried downslope. The shallow water-repellent layer can last for more than a year after a wildfire, keeping rain from soaking into the ground and creating a surface layer more vulnerable to landslides or debris flows. Sediment movement off hillslopes can increase by a factor of three hundred during the year immediately after an intense fire.

Small, frequent fires are not so effective at causing hillslopes to fail. The smaller fires are less likely to kill trees and shrubs, and the roots of these still-living plants help anchor the soil and maintain its ability to soak up water. More frequent fires create smaller debris flows and landslides, preventing the accumulation of thick layers of sediment that can fail in large debris flows. Frequent fires also remove built-up fuel, making the occurrence of an inferno less likely. In essence, small, frequent fires promote landscape stability.

Smaller fires do not prevent the infernos from occurring, however. Geological and botanical records covering thousands of years in various parts of the western United States indicate that periodic large fires occur no matter what. The frequency of these fires differs from region to region. In the Sierra Nevada, scars on giant sequoia trees record large fires that have replaced whole stands of lesser trees every few decades. In the Front Range of Colorado, higher intensity crown fires have burned the spruce and fir forests once every three hundred to four hundred years. These more intense fires cause periods of substantial landscape change because they mobilize large volumes of sediment, stripping hillslopes to bedrock that then begins to weather slowly into sediment. Regions such as the Front Range have likely experienced episodic landscape and ecological change for millennia, with hundreds of years of relative stability interrupted by short periods of dramatic change.

.   .   .

People probably changed the wildfire regime of North America as soon as they reached the continent. Archeological and historical evidence indicates that Native Americans from Florida to Washington used fire as a tool to shape the landscape. Fire promoted the growth of plants on which deer liked to graze. Fire helped to clear plots for crops. Fire drove game animals before it, concentrating and trapping the animals where hunters could reach them. And fire cleared the ground for favored plants that could be gathered for food and medicines. During thousands of years of human occupation of North America, many plant and animal communities evolved and adapted to frequent smaller fires started by humans.

The first Europeans to reach North America generally did not recognize the role that fires set by humans had played in shaping the landscape. The Europeans assumed that the Native American peoples were too primitive to have actively shaped their surroundings. Because the Europeans thought the landscape was a pristine wilderness, they did little to help start fires. They used fire on a limited basis to clear plots for homes and crops, but they did not deliberately set ground fires to clear underbrush from extensive tracts of forest.

As fire frequency decreased, fuel materials built up in many forest environments. A combination of climatic conditions and fuel accumulation led to a series of firestorms during the latter part of the nineteenth century, as reviewed in Stephen Pyne's fascinating book *Fire in America*. These enormous, fast-moving, intense fires were sometimes set by sparks from coal-burning locomotives, sometimes by agricultural fires that escaped. Some of the firestorms burned for months, destroying extensive areas of forest and settlement in the eastern and midwestern United States.

The terror inspired by these seemingly unstoppable catastrophes, along with the steady decrease in forest cover and timber supplies in the United States, and the associated problems with erosion, led to a general policy of fire suppression. Many mountainous regions in the western United States experienced massive deforestation during the second half of the 1800s as discoveries of precious metals led to sudden population increases and the destruction of standing trees as a result of both overharvest and forest fires started by sparks from trains or campfires. As the denuded mountain hillslopes hemorrhaged sediment, agricultural irrigators settling at the base of the mountains found that their intake structures were being destroyed by siltation. They called for a halt to deforestation, and Congress established the national forest reserves and the first national parks.

Some of the early rangers for the Forest Service and the National Park Service were recruited from the army, and their policy was that all fires be put out immediately. Fire represented wanton destruction of precious natural resources. Popular culture represented wildfire as evil, as in the forest-fire scene in Walt Disney's animated feature *Bambi*. So did the language used to describe *fighting* fires. Those who stopped the depredations of fire were heroes.

As fires became less frequent, fuel again accumulated in the forests. Plants that required fire to complete their life cycles became less common, as did the animals that relied on the plants. In some cases plant and animal species became so rare that they were listed as threatened or endangered. And when a wildfire did break out, it was more likely to be a violent, uncontrollable burn that killed trees rather than merely clearing out the forest understory.

By the 1980s, natural resource managers gradually began to espouse a controlled burn policy, so that smaller, frequent fires similar to those that the Native Americans had employed would once again be used to promote forest health and diversity. Controlled burns have remained controversial, however, among a public educated for decades by messages promoted via Smokey Bear (Only You Can Prevent Forest Fires!). For starters, controlled burns can escape control. As the economy and demographics of the so-called New West shift toward increasing numbers of people living in smaller communities dispersed throughout forested lands, residents who choose to live near the forest for the scenery become very unhappy at the thought of a wildfire burning either the surroundings or their homes. And there remain many people who see wildfires as nothing but the unfortunate destruction of timber supplies or wildlands. Those who reject the loss of timber to a fire see forests as cash reserves, with wildfires playing the role of bank robbers. Those who lament the destruction of nature by wildfires seem to regard a forested landscape as a "scene" or a painting, rather than a dynamic system. The fire ruins the view for decades.

After three years of severe drought, Colorado experienced a summer of large and intense wildfires during 2002. The Hayman fire alone burned 137,000 acres, the largest fire to date in Colorado's history. People living in the mountains assumed a siege mentality. They cleared low branches, snags, grasses, and brush from around their houses. They scanned the skies anxiously for rain and relied on firefighters to contain and suppress fires, as well as to protect homes and evacuate residents. When a fire occurred, the residents thanked the firefighters with homemade signs and with cookies and coffee. When a fire destroyed a home, the residents mourned dreams and memories lost and often vowed to rebuild. Media coverage focused on the physically demanding labor of the firefighters, on the strain of working for days in hot, dirty conditions while carrying heavy equipment and getting little sleep. Fire was definitely the enemy, but human courage and endurance were pitted against it as neighbor helped neighbor and firefighters risked their lives.

The courage and commitment displayed by many individuals in Colorado during the summer of 2002 were impressive. But very few people seemed to be questioning why such courage and commitment were needed. What I missed in the media coverage was an intelligent examination of how American policies of historical fire suppression, as well as minimal restriction on the development of private lands and maximum expectations of federal assistance for disaster relief, had allowed the spread of valuable, flammable homes in forested lands. Periodic wildfires will only become more intense in this region as a result of fire suppression. In 2002, an estimated one million people lived in Colorado's red zone, the aptly named interface between wildlands and human dwellings: a 30 percent increase in population within a decade. The number is expected to double by the year 2020. The federal fire budget for the whole country rose from $1.8 billion in 2000 to $2.9 billion in 2001 and $2.3 billion in 2002.

It seems to me that these costs will continue to increase so long as various levels of government allow people to build homes throughout forested regions that were much more sparsely inhabited only a decade ago. Population in the mountain West grew by 25 percent during the 1990s, making it the fastest-growing region in the country. Many of the newcomers have moved to Denver and Salt Lake City, but rural counties are growing even faster. As dispersed communities become more widespread, land managers get trapped between a rock and a hard place: Residents mistrust controlled burns, but they truly fear intense fires and expect every effort to be made to stop such fires. And once the fires are put out, the public increasingly expects land managers to also control the aftereffects: the flash floods and debris flows. The Forest Service spent an estimated $50 million fighting wildfires in the Colorado Front Range during 2002, and an additional $24 million on rehabilitation of burned lands following the fires. Rehabilitation following the Hayman fire accounted for $17 million of the total.

. . .

Does rehabilitation really change water and sediment movement from hillslopes following a fire, or is it a placebo that either discourages the public from suing the Forest Service or makes people feel good because they are doing something? There has been surprisingly little monitoring of the effectiveness of rehabilitation, but a Forest Service review following the Hayman fire characterized overall rehabilitation effectiveness as poor. Rehabilitation techniques usually entail trying to stabilize burned slopes by seeding them with quick-growing grasses or mulching them with hay or other materials that protect the surface from rainfall and runoff. It is crucial that any mulch used be weed free. Certified weed-free hay shipped in from Kansas following the summer 2001 fires in Colorado turned out to be contaminated with invasive cheatgrass. Some of the areas treated with the hay now have impressive stands of cheatgrass, which excludes native grasses and increases the likelihood that wildfires will start and spread in the future, because this flammable grass dries relatively early in summer.

Hillslopes under rehabilitation may be worked with machines that roughen the surface by digging trenches across the slope or ripping small pits in it, so that water and sediment moving downhill are trapped. Barriers such as logs or hay bales placed across the slope are used to trap water and sediment moving downhill. As with mulching, details make the difference. One of my students working for the Forest Service told me of a contractor who filled his crew with homeless people from Denver. They put the log barriers parallel to the slopes, rather than perpendicular, very effectively channeling water and sediment downslope.

Hillslopes under rehabilitation are also sometimes treated with chemicals that cause soil particles to clump together. Some of these chemicals can be toxic to fish, which creates problems when the chemicals eventually flush downslope into streams.

My colleague Lee MacDonald and several of his students conducted extensive monitoring following the Hayman fire. His work and other studies of burned areas suggest that rehabilitation measures make little difference in the amount of sediment eroded from burned slopes during the years immediately following a fire, especially if intense rains occur within the first year. Repeated applications of mulch seem to be the only treatment that can reduce sediment movement, but this treatment becomes progressively less effective against larger storms. The massive amounts of sediment available to be eroded from hillslopes in areas such as the Buffalo Creek watershed simply overwhelm any rehabilitation efforts. Once vegetation begins to naturally reestablish, which it does within a year, sediment movement begins to decline. Sediment movement from even severely burned areas usually approaches preburn levels within three years, but up to ten years may be necessary for ground cover to completely regrow and sediment movement to return to preburn levels.

Rehabilitation measures also focus on stabilizing river channels as the sediment eroded from burned slopes overwhelms the channels. Rapid channel change can destroy fish habitat or result in worse flooding downstream from burned areas. To reduce these effects, managers build small structures from hay bales or plastic sheeting to slow water and trap sediment before it enters streams. They bolt logs or rocks into the streambed or banks to create fixed points that create a scouring effect and maintain pools. They strengthen banks with rocks, logs, or hay bales and dredge channels to remove excess sediment.

As with the hillslope rehabilitation measures, these techniques have questionable success. The relatively few studies that have been done suggest that stream rehabilitation following a forest fire may help for the first few months if the fire has not been so severe that all vegetation is removed. But stream rehabilitation is unlikely to effect much change over longer time periods, because the stream often becomes so unstable that it erodes the rehabilitation structures. Within a decade of the fire, streams that received rehabilitation measures are not in demonstrably better condition than streams that did not.

My experience with rehabilitation measures suggests that the effort and expense are wasted. Hillslopes and streams largely become stable on their own schedules, whether or not rehabilitation is attempted. Rehabilitation attempts may be necessary to discourage disgruntled property owners from suing the federal government for damages following a wildfire, but in many cases that is about all they can accomplish.

Because wildfires and their aftereffects cannot always be controlled, public attention also focuses on preventing fires. Even though controlled burns have a bad reputation at present among some groups, controlled fires are still widely used for managing federal lands. Thinning has also been proposed, whether selective cutting of trees and brush, or clear-cuts of scattered patches. *Thinning* may also be a euphemism for clear-cutting fairly large areas of forest. In theory, thinning removes enough of the accumulated fuel in a forest to prevent intense crown fires. It can reduce fuel loads provided all slash piles are removed or burned. But thinning does not mimic a forest fire.

Forest fires kill plants and animals. I have yet to see anything that looks quite as devastated as the blackened, smoking remains of a formerly cool, green forest immediately after a hot fire. Yet fire does not permanently devastate the forest. The heat of the flames releases plant seeds that remain dormant until they are exposed to this heat. Nutrients in the ash return to the soil. Sunlight flooding onto the newly cleared ground favors pioneer plant species that require abundant light and little competition from other plants. The pioneers in turn provide shade that allows other plant species to germinate. Birds feast on the insects flushed into the air by the smoke and exposed on the bare ground. Scavengers from microbes to beetles to vultures eat the animals trapped and killed by the fire. Nutrients present in the forest before the fire are redistributed by the fire, and new communities swiftly replace those destroyed in the fire. Fire-killed trees provide nesting sites for birds. The forest is enriched by fire.

Thinning, by contrast, removes nutrients. Roads must be built to allow heavy equipment such as logging trucks into the site. In otherwise forested areas, unpaved roads form lines along which the landscape unravels. Compacted road surfaces interrupt the downslope movement of water and sediment, triggering landslides that dump large amounts of sediment into streams. The roads themselves erode into "washboard" surfaces and gullies, leaking sediment into streams with every rain. Trucks and other equipment compact the soil where they travel off-road, reducing the ability of the soil to soak up water and increasing sediment erosion. These effects are similar to those induced by wildfires but persist longer. Nutrients stored in each harvested log are removed from the forest during thinning, rather than being left in place for fungi and microbes and ants to recycle into the forest soil. Habitat critical to bird species that nest in rotting snags or to fungi and insects that live in rotting logs is lost. Depending on how thinning is conducted, it may not create the mosaic of different habitat types that form following fire. These different habitat types favor diverse plant and animal species, creating a healthier forest community more resilient to disturbances in the form of small fires, windstorms, droughts, or severe winters.

.   .   .

The debates surrounding wildfires and forest management are similar to those surrounding the rehabilitation and management of rivers. What are we managing toward? What conditions would we like to exist in the forest or the river? This question is unlikely to be answered any time soon because of the diversity of opinions present in American society. Are the forests and the rivers commercial commodities on which humans can draw at will? Or functioning ecosystems with a fundamental right to exist apart from human convenience? Or critical life-support systems on which humans depend for existence? Or all of these? American natural resource policy is likely to continue its historically meandering course as long as diametrically opposed, strongly held views exist on what a forest or a river *should be*.

Land management is equally difficult to address: Given a desired outcome, how should we get there? What actions should be taken to ensure that forests produce maximum board-feet, or maximum biodiversity, or maximum erosion control? How should we manage rivers to ensure minimal flooding, maximum water supply or quality, and optimal fish habitat? Answers to these questions rest more on technical expertise but still require ethical judgments. Because forests and rivers constantly change at timescales of years to millennia, technical knowledge is not completely sufficient to answer questions of management technique. The experiments that would show how a forest responds over a century to different management strategies take too long to run fully, and we must extrapolate from shorter, incomplete experiments.

For residents of the West, questions of land management inspire not cool, intellectual arguments but rather heated emotional ones in which all sides seem to see everything they value as being at stake. The arguments might not be so emotionally fraught were it not for the stories that Americans have told themselves in the past. Would governmental restriction on individual or corporate rights to mine resources matter so much if Americans had not believed for generations that we are a nation of pioneers destined to subdue the wilderness, and a nation of entrepreneurs, where individual freedom to maximize profit is sacred? Would the preservation of wilderness matter so much if we did not believe that our immigrant ancestors found a pristine wilderness—a new world—which it is our duty to save from commercialization?

I know that my own opinions are heavily influenced by the importance that wilderness and a self-regulating natural world hold for me. My growing awareness of the long history of human impacts in the United States has been difficult, because it has forced me to acknowledge that pristine wilderness does not exist. From deliberately setting fires to hunting the largest game animals to extinction, the first humans to reach this continent began thousands of years ago to alter the ecosystems they encountered. But contemporary commercial and industrial society has stepped up the pace and intensity of those alterations dramatically.

.   .   .

Walking the slopes near Buffalo Creek after the fire, I came across a flattened, thoroughly dead mouse belly up on the ground among the standing blackened remains of trees. Patterned black-

and-orange beetles worked over the carcass, and a smaller beetle with a textured black shell squirmed out from beneath the body. The carcass had no smell. The mouse's mouth was open, its front paws raised in a gesture like surrender. Occasionally the energetic beetles caused the whole carcass to squirm. I found it disquieting to see flies crawling over the delicate nose and whiskers that could no longer twitch to send them away. The beetles nudged the carcass against a boulder of weathered pink granite. Sooty black lichens curled like flakes of ash grew near its base.

These burned slopes that looked so lifeless from a distance were, on close view, still full of the processes by which one organism transfers matter to another. The whole drainage basin—the whole process of landscape development—constitutes a cycling of matter and energy, from the decay of radioactive elements powering the tectonic activity that created the Rockies, to the thermal energy of the sun drawing water from the oceans and raining it down to shape drainage basins, to the energy stored in plants and released as fire, and finally, to the energy transferred from sun to plant to mouse tissue that is now being reclaimed by the beetles. Fire is a part of this constant change and cycling, and some form of fire will occur no matter what management strategies are employed. If we could remember this while admiring the magnificent scenery, we might develop an intuitive understanding of landscape processes and of our own place among them. We might make our peace with fire.

Let It Snow!

Halloween is upon us, and in Colorado it usually brings snow. The weather at the eastern base of the Rockies is only broadly predictable. Summer is warmer than winter, and spring is windier than any other season. Beyond these predictions it is dangerous to stray, with one exception: it will snow on Halloween. Inevitably, excited children dressed in costumes lumpy over warm clothing will rush from door to door amid wind-whipped snow flurries.

The steadily shortening hours of sunlight bring out hues of gold, orange, and red in the leaves of the maples and cottonwoods in town. Up in the mountains, aspen set the somber green pine forests aglow with vivid patches of gold and bronze. A coating of frost evades the warmth of the rising sun each morning, and wood smoke scents the air. I imagine a hint of snow in the air. The nesting urge grows strong, so I spend long evenings reading and writing beside the fireplace.

As I settle down, I feel that winter should be equally settled. The Halloween snow should be followed by heavier, more persistent snowfalls, and the world should assume hues of white and gray. Undoubtedly this is the product of growing up in northern Ohio. My memory is notoriously fallible, but I remember white winters that took their time before melting into spring. Winters on the Colorado plains do not quite meet my expectations. Snow falls on Halloween, but it melts within a day or two, and temperatures are likely to return to the midsixties or even seventies during the daytime. Snowfalls throughout the winter are succeeded by bright, snow-melting days, and winter here is most distinguished by a mildly cold dryness and a thinner, paler quality to the sunlight.

Temperatures in the mountains are of course colder, and the snow more persistent, but even in the mountains the snowpack can be unreliable. During drought years, the shrubby willows in my favorite skiing meadow rise well above the snow. During good years, only a slight undulation in the prevailing whiteness hints at their submerged presence.

Drought is hardly a once-in-a-lifetime occurrence in Colorado. Historical records indicate that, most of the time, at least 5 percent of the state is suffering droughts of three months' to two years' duration. This poses a challenge for living organisms. Plants and animals native to various parts of the state evolved adaptations to cope with drought. The first humans to live here were nomads who walked onward when drought struck. As people of European descent immigrated to Colorado, drought became harder to cope with because we Europeans like to settle down. Early agriculturalists who planted thirsty crops imported from wetter regions developed an elaborate water-engineering system to provide water when the rains didn't fall. Urban communities adopted this water-engineering strategy.

Winter at Montgomery Pass, Poudre Canyon, Colorado

One of the attractive qualities of drought-prone regions is that they have plenty of sunshine. In Colorado, the sun shines on lovely scenery. This combination supports a tourism industry that annually injects around nine billion dollars into the state's economy. Much of this tourism relies on water. Anglers, boaters, and skiers in particular can hardly pursue their activities in the absence of water, and Colorado tourism is driven primarily by the ski resorts and associated infrastructure. Skiers purchased 2.3 million fewer lift tickets during the 1977 drought than they had the year before, a pattern repeated during a drought in the early 1980s.

This sort of yearly fluctuation does not make ski industrialists happy. The phrase *ski industrialists* sounds awkward, but it provides the most apt description of contemporary ski resorts in

Colorado. Decades ago, somebody had the bright idea of putting ski lifts in mountainous areas. An easy ride to the top of a mountain made it more fun to zoom back down on skis, and the addition of a fire-warmed chalet where a skier could rest with a hot drink made the activity even more attractive. But these operations did not generate huge profits for the owners, and in drought years the profit margins shrank even lower. Then somebody got the idea of building small resort communities to support the activity of skiing. Today a lift ticket in Colorado sells for around seventy dollars and a season pass runs three hundred dollars, but the major profits come from the condominiums, hotels, restaurants, and shops accompanying the ski runs. The runs themselves are hardly wilderness adventures. Ski industrialists invest millions of dollars in grading and shaping the runs, removing boulders and tree stumps, putting in lifts and maintenance roads, and laying underground pipes for snowmaking.

Summit County encompasses many of the major ski areas in Colorado. Population in the county nearly doubled between 1990 and 2000, driven in part by a doubling in skier visits between 1980 and 2000. These twenty years coincided with Colorado's second-longest sustained wet period in recorded history and the most drought-free period since 1890. As the state has experienced severe drought conditions since 2000, skier visits have dropped. Ski industrialists have responded by stepping up the artificial addition of snow to ski slopes—snowmaking—and by expanding their operations. The last response might seem counterintuitive, but the hope is that, in a competitive economy, the largest and most lavish operation wins. Part of this expansion is to build a larger ski village with a run behind every condo or hotel.

Ski resorts in Colorado largely operate on national forestlands.

Ski owners receive a forty-year permit to use the land and in return pay 4 percent or less of their profits to the federal government. One result of this arrangement is that ski industrialists, who legally rent their lands at the pleasure of the government, in fact dictate policies to the national forests. Why? Because they inject nine billion dollars annually into the state economy, and because the forest administrators seeking a new public mandate in the face of declining timber sales have chosen to boost recreation.

The U.S. Forest Service has begun to question the wisdom of ski-industry expansion on forestlands because of concerns regarding the sources of the water used in snowmaking and the subsequent effects of the snow once it melts. Existing ski facilities require a lot of water for snowmaking, and expanded facilities will of course require even more. Most resorts begin snowmaking in mid-October, two weeks before we on the plains are settling down to enjoy the first brief snow flurries on Halloween. Colorado's Front Range resorts race each other to be the first to open for the season. The winner makes national news and draws reservations from the profitable overnight visitors.

The water supply for snowmaking is drawn from nearby streams. These streams are usually already heavily allocated, or drawn upon, for consumptive, off-channel uses such as drinking-water provisioning and agricultural irrigation, either in the mountains or at lower elevations. The streams can also be contaminated as a result of past human activities in the mountains. Colorado contains more than seven thousand abandoned mine sites, mostly dating from the nineteenth century. Many of these mines contaminate surface waters with acidic, metal-enriched wastes still leaching from the mine tunnels and tailings. Contaminated waters pose a threat to all life-forms, from algae to humans, that come in contact with the waters. As ski resorts draw more heavily from mountain streams, the use of these contaminated waters increases. And when the now-contaminated snow melts, the meltwater runs into adjacent uncontaminated streams and spreads the poisons downstream over greater areas. The "white gold" of snow is now worth more than the placer gold that miners tore from the mountain streambeds, but the legacy of mining continues to affect water today.

The electrical energy necessary to run snowmaking air compressors and water pumps comes primarily from coal-burning power plants such as those at the cities of Craig and Hayden. A federal court found that the Hayden plant violated clean air standards at least nineteen thousand times in five years during the mid-1990s. U.S. Geological Survey scientists found that snow and rain acidified by sulfur dioxide produced at Craig and Hayden has crippled or killed 40 to 100 percent of the amphibians in ponds of the Mount Zirkel Wilderness in northern Colorado, which is downwind from the two power plants. In his thoroughly researched book *Downhill Slide,* Hal Clifford estimates that a ski area paying a million dollars for electricity to run its snowmaking compressor during a typical ski season pays a utility to burn fourteen million tons of coal that in turn produces thirty million tons of airborne carbon dioxide, a major greenhouse gas.

Melting, artificially created snow has other effects on small stream channels draining the ski slopes. Colorado's White River National Forest includes several major ski areas that together sell twelve to fourteen million lift tickets each year. Because the ski industrialists located within this forest seek to expand, forest hydrologists decided to systematically examine the effects of snowmaking on these stream channels.

Snowmaking can double the amount of water that, during peak snowmelt, runs into adjacent streams. Higher flows carve into the stream channel, creating deep gullies in what had been small swales with flow only during snowmelt. Larger streams that flow throughout the year can receive so much extra water that the banks become undercut and adjacent trees topple into the channel. The sediment mobilized by this enhanced erosion is deposited downstream in the larger channels, where it fills pools and reduces fish habitat. Unpaved maintenance roads switch-backing across the ski slopes can also be a source of substantial sediment wherever they cross a stream. Some of these effects can be mitigated by moving ski runs away from existing headwater channels, but the ski operators prefer to stabilize the channels by lining them with large rocks or landscaping fabric, or by building small dams to trap sediment. Such mitigation strategies seldom work well. If a stream is eroding because of increased flow, or it is accumulating excess sediment, the only effective way to mitigate these effects is to reduce the flow or remove the source of excess sediment.

The forest plan for the White River National Forest, which guides management decisions, specifies that the Forest Service must seek to conserve stream conditions and aquatic habitat in perennial streams. Forest hydrologists judge the condition of affected streams by comparing them to unimpacted reference streams. Conditions that vary within 10 percent—for example, channel width plus or minus 10 percent of the width of an analogous reference stream—constitute the robust health that forest planners seek to maintain. This criterion sounds straightforward, but can be very difficult to implement because reference streams do not provide a uniform standard for comparison.

The width of a stream channel at any given point reflects its upstream drainage area and the amount of flow, the rock type underlying the stream valley, the history of the valley, the width of the valley bottom, processes occurring on the adjacent hillslopes, the amount and distribution of logs in the channel, and on and on through a long list of potential controls. The shape of a mountain stream typically varies substantially in a downstream direction. It might have a wide, shallow channel, for example, where it flows through a broad valley once occupied by glaciers. As the stream passes over the terminal moraine that marks the downslope end of the long-gone glacier, the stream channel abruptly grows steeper and narrower, and it flows in whitewater cascades over huge boulders. Below the moraine the stream once again meanders across a broader valley until it encounters a large debris fan where a landslide came down the valley side slope, constricting the stream channel and pushing it toward the opposite side of the valley. Which of these distinctly different stream segments should provide reference conditions for channels influenced by snowmaking? The usual response to this dilemma is to gather measurements from a sufficiently large number of reference and impacted streams to allow statistical comparisons. Any given pair of streams could be chosen to demonstrate large differences between a reference and an impacted stream, or no differences. The intent in using large data sets is to seek out consistent differences.

My graduate student Gabrielle David worked with hydrologists from the White River National Forest to define the range of variability in stream characteristics that might respond to snowmaking. She spent two summers carefully measuring all the variables that could influence channel characteristics, as well as the variables that could reflect the influence of snowmaking activities, for two dozen reference and two dozen impacted streams. Her data included well over a hundred variables for each stream, which she then patiently examined using statistical techniques that would help her to understand how the combined variables connected to patterns observed among the streams.

One of Gabrielle's findings was that stream bank vegetation, and particularly the presence of shrubby willows, substantially increases resistance to erosion that might otherwise result from the increased flows coming from artificially produced snow as it melts. The presence of willows is thus part of the story in explaining why some streams erode and become unstable in response to snowmaking activities and some do not, but willows explain only part of the observed variability. The underlying rock type is another important factor: streams flowing over granitic rocks in the White River National Forest are more likely to show differences between reference and impacted streams than are streams flowing over sedimentary rocks. This likely reflects the fact that the sedimentary rocks are less resistant to the processes of weathering that decompose rock into sediment, and to the processes of erosion that redistribute the sediment. Streams flowing over sedimentary rocks are highly prone to avalanches, landslides, and debris flows, even in the absence of any land use such as snowmaking or an increased density of roads. It is difficult to detect the effects of land use when the streams are naturally so unstable.

When the reference and impacted streams are plotted on an x-y diagram in which the two axes represent a combination of the characteristics of each stream, such as drainage area and streambed slope, the reference and impacted streams form two clusters of data. Drawing a circle around each cluster results in two circles with a large amount of overlap. In other words, there are many reference streams with characteristics—such as width-depth ratio or bank stability—that are very similar to those of impacted streams, and vice versa.

I have seen this degree of overlap presented repeatedly in studies that attempt to distinguish streams altered by land use activities from more natural, reference streams. The extreme cases are easy to identify: streams where complete deforestation of the drainage basin results in massive increases in sediment entering the stream, for example. But the great majority of the more moderate effects of land use are difficult to discern among groups of streams, because any particular stream or segment of a stream reflects multiple, interacting controls.

My work on the effects of skiing on streams in the White River National Forest revealed consequences that I had not thought of in connection with commercial skiing. These new perceptions make it difficult to welcome the first snowfall of the year with the same innocent pleasure. Now the snow represents not only a fresh change of season or a new year but also arrives burdened with the intricacies of water politics and economics.

Equifinality

I used to have lunch with some elderly colleagues who had retired from my department at the university. One day one of them who had cleaned out his basement brought me a stack of books, wondering if I wanted any of them. Looking through the stack, I came across copies of the original editions of *Report on the Geology of the Henry Mountains* (1877) and *Lake Bonneville* (1890), both by G.K. Gilbert of the U.S. Geological Survey. Gilbert is now considered the father of modern American geomorphology. (Like many sciences, geomorphology was born without a mother.)

I eagerly accepted the books, savoring the leather bindings darkened with age. Inside the cover of the Henry Mountains volume is an elegant inscription that it took me some time to decipher: *Geo. Bird Grinnell, October 20, 1879, New Haven, Connecticut.* I could only wonder at the peregrinations by which a book owned by George Grinnell found its way into the basement of Joe Weiss in Fort Collins. Grinnell was a leading biologist and anthropologist who served as a naturalist on expeditions to the

West in the 1870s, and who founded the first Audubon Society. He also accompanied Gilbert, John Muir, John Burroughs, and others on the 1899 expedition to Alaska organized by railroad baron Edward H. Harriman. Harriman filled his private yacht with scientists and naturalists and went off to see the sights in Glacier Bay and other points along the Alaskan coast, providing an otherwise unattainable, if sometimes frustratingly limited, opportunity to expedition members.

I gave the Gilbert books pride of place in my library, putting them on the top shelf, where a deep overhang protected them from direct sunlight. The smooth, heavy pages of the books have weathered to a dull golden orange along their edges. Turning the pages releases the smell of old books, a smell of libraries and of hours spent in thought. These are technical monographs, but they have a conversational tone absent from modern scientific writing. Gilbert's pen-and-ink sketch of a mule's head, bearing the caption "Ways and means," accompanies a section titled "How to Reach the Henry Mountains."

Gilbert is considered the father of modern geomorphology because he pioneered the quantitative and experimental approaches widely favored today, inferring processes of landscape development from his study of mathematical relations among physical forces. When I seek to understand the response of a river channel to a debris flow, I measure rather than just observe, and I measure a sufficiently large collection of variables to perform statistical analyses or to calculate forces. I emphasize quantifiable evidence because I have been trained to think that only "hard" numerical evidence, which can presumably be duplicated by another investigator elsewhere, will help me overcome any inherent biases I might have and, thus, demonstrate to my colleagues the appropriateness of my interpretations. More than any other geologist of his era, Gilbert started us down this path. He solved riddles such as the volcanic origin of Utah's Henry Mountains by using analogies to mechanics and mathematics. When he investigated the effects of placer mining in California's Sierra Nevada, he built experimental flumes to study the movement of sediment downstream. Gilbert defined the methods to be used in field geology during what his biographer, Stephen Pyne, called the heroic age of American geology.

During the latter half of the nineteenth century, geologists such as Gilbert and his boss at the Geological Survey, John Wesley Powell, undertook heroic journeys to reach and study their field sites. Powell risked his life floating down the unmapped Colorado River. Colleagues such as Ferdinand Hayden set out to survey the West's natural resources during a time when the Native Americans were trying to keep Europeans from usurping their lands. As John McPhee wrote, "Professor Hayden was accorded the special status that all benevolent people reserve for the mentally disadvantaged. . . . The Sioux named him He Who Picks Up Rocks Running, and to all hostilities thereafter Hayden remained immune." McPhee describes the charismatic Clarence King expressing "immense enjoyment" at his survey of the Stillwater Ranges, despite a lightning strike that turned the right half of his body brown for days.

The knowledge gained by the pioneering geologists was integrated into the heroic American pursuit of identifying and exploiting natural resources. The men developing the knowledge were not always treated as heroes, however. I have thought of Powell, in particular, when I fly over the West. Viewing the Earth's surface from an airplane provides a chance to step back to an older approach to landscape, unconstrained by political boundaries. At thirty thousand feet, the landscape takes on abstract patterns of topography, vegetation, and land use. I can see the reality of the land beyond political boundaries, which in the West are arbitrary, ruler-straight lines slapped onto a map. The eastern states grew more naturally, and their winding boundaries largely reflect the rivers and coasts that define the land. The western state boundaries reflect bureaucratic ignorance and standardization without regard for topography or culture. Thus we have the Cubist designs of Colorado, Wyoming, New Mexico, and Utah, for example, isolating the Mormons of the northern Arizona Strip from their hub in Salt Lake, and the Hispanics of southern Colorado from their roots along the central Rio Grande, as well as truncating watersheds and creating water wars between states.

Powell proposed settling the land by watersheds and planning settlement for the common good rather than allowing individualism to govern settlement. His 1878 *Report on the Lands of the Arid Region of the United States* presented an alternative to the land-office grid pattern of 160 acres per settler, which was designed to produce the Jeffersonian ideal of a nation of yeomen farmers. The design may have worked for wetter regions, but from Powell's observations in the arid West he knew that more acres were required for most land uses, and that the number of acres had to be flexible to account for different qualities of land as well as for different activities, such as raising cattle versus raising wheat. Powell preached water-imposed population limits and proposed that no further land be granted for irrigation. His proposals brought howls of outrage from western boosters unwilling to admit that such limits existed, let alone that they had already been reached. Powell's ideas were never taken seriously by congressmen who supported the idea of corporations or wealthy individuals making fortunes on western land speculation and government subsidies, or by congressmen who believed that the West could be as well-watered and densely settled as the East. They buried Powell's vision of the West's future, then robbed him of his directorship of the Geological Survey.

Despite the shortsightedness of politicians, the golden age of American geomorphology was also heroic in the speed with which understanding of landforms advanced. Back in the 1780s, Scotland's James Hutton had proposed that relatively small rivers could form great valleys by slowly and inexorably cutting downward year after year. In the 1840s, Louis Agassiz of Switzerland complicated matters by suggesting that some great valleys had actually been carved by glaciers rather than by rivers. Geologists were hotly debating the role of rivers in shaping regional landscapes when, after the Civil War, American geomorphologists began to explore the West. Powell ran the rapids of the mighty Colorado River through the Grand Canyon, where no glacier had been. He followed the Green River as it cut like a knife through Utah's Split Mountain rather than taking the path of least resistance around the mountain.

During this journey, Powell studied the sedimentary records preserved in rocks of rivers that had cut their way down to a new level, remained stable, then cut down again. From these observations, he and his colleagues developed ideas of base level, antecedent drainage, and superimposed rivers. They recognized that a river will not cut down below its downstream-most point (base level), but that it can cut downward at a rate equal to the uplift of a mountain that is being raised in its path (antecedent drainage), or cut through a buried mountain range that is uncovered as the river cuts downward (superimposed). These ideas allowed the early geologists to explain the wild landscapes of the Colorado Plateau, where a high-elevation surface, pancake flat, suddenly drops off into a gorge containing a muddy, turbulent river.

Geologists could not really conceive of the rapid incision of a river through a rising landmass until they explored the Colorado Plateau, where the land is moving up and the rivers are cutting down to keep pace, creating immense features like the Grand Canyon in a few million years—the blink of a geological eye. The next generation of geologists could not explain the movements of the Earth's crust in tectonic plates, despite the mounting evidence of matching fossils in Antarctica and Australia, or glacial deposits in tropical latitudes, until the mapping of the Atlantic Ocean floor after World War II revealed a great north-south rift along which new seafloor is created.

Technology and experience condition one's esthetic and spiritual, as well as scientific, perceptions of the world. Think of the fragile blue planet photographed by astronauts, or the checkerboard of the American plains revealed by an airplane flight over the region. Esthetic and spiritual perceptions—a culture's worldview—in turn influence scientific inquiry. Most scientists only perceive what they conceive of as possible. Natural philosophers living in Britain during the seventeenth century tried to develop explanations for the appearance of the Earth's surface. Because they assumed that the Earth was only a few thousand years old, as implied in the Bible, and because they were not field scientists or global travelers who had directly observed volcanic eruptions or large floods, they worked from the premise that the Earth's surface had been initially created during some catastrophe such as Noah's flood and then had remained static ever since. They did not conceive of the possibility that the Earth was much older, or that it had been continually changing since its formation.

Ideally, scientists operate under the principle of multiple working hypotheses, a concept formally developed by Gilbert's contemporary Thomas C. Chamberlin. This concept proposes that investigators simultaneously consider many alternative explanations for observed natural processes. Most of us find it difficult, however, to really consider more than a few potential explanations. We are usually left with one or two hypotheses that are constrained by our worldview.

Part of the worldview held by Americans in the late nineteenth century was that humans could severely alter and degrade the natural environment. The rapid cutting of forests and overhunting of many species of animals, birds, and fish led some Americans to realize that the seemingly unlimited abundance of the continent did in fact have limits. Men such as Grinnell, John Muir, Theodore Roosevelt, and Gifford Pinchot began to articulate a national conservation ethic that involved governmental limits on resource use. These ideas influenced the geomorphologists who began to investigate the formation of arroyos in the southwestern United States in the first years of the twentieth century.

Anglo settlers in the dry portions of the Intermountain West enthusiastically stocked the grasslands with cattle, horses, and sheep. These grasslands had evolved with bison, pronghorn antelope, mule deer, and other grazers, but the intensity and style of grazing differed between the domestic and wild animals. The shortgrass prairie of the Western drylands is a conservative ecosystem, tightly recycling the limited available nutrients and water and prone to both erosion and loss of soil fertility if something removes the nutritious bunchgrasses. The huge cattle drives and intensive stocking of public rangelands after the Civil War effectively removed the grasses and shrubs from large regions of the West.

Dense concentrations of grazing animals ate the surface portions of plants. The animals' hooves compacted the ground surface, reducing the ability of the soil to absorb rainfall and snowmelt. As the soil grew drier, it was less able to support plants. Precipitation falling on the hardened ground was more likely to run quickly downslope, taking with it the fertile layer of topsoil and creating rills that eroded to become gullies. The water and soil rapidly moving down the hillsides became flash floods in the larger channels, which caused more erosion and deposited sediment downstream.

Many of the stream channels in the drier parts of the West cut dramatic steep-walled gullies, called arroyos, between approximately 1860 and 1940. Nineteenth-century photographs of Kanab Creek in southern Utah show a flat, unchanneled valley with broad, moist swales and dense grasses. By 1935, a channel 75 feet deep and 360 feet wide divided the meadow. Fifty years later, the gash had been filled once more, leaving only a slight depression along the stream channel.

Incision by streams had damaged valuable crop and rangelands and caused sediment deposition and exacerbated flooding downstream, catching the attention of scientists and the public. The obvious explanation was that overgrazing had triggered the stream incision. Resource managers began to establish maximum limits for grazing stock on public lands, looking to scientists to help them define these limits.

Some of the geomorphologists studying arroyos suggested that climatic variability could play a role in stream incision. They

Arroyos at Pawnee National Grassland, Colorado

noted that sedimentary deposits exposed in the banks of the eroding channels indicated that channels had been incised and refilled many times during the past two million years, long before domestic grazing animals were present. Drought could cause stream cutting, by killing off vegetation and promoting rapid runoff from hillslopes into streams when rain did fall. Or a wetter climate could cause stream cutting by producing more rainfall and, with it, runoff that eroded streams. Or even when average rainfall remained the same, intense rains such as thunderstorms could produce floods that incised streams. The debate remained lively for decades as geomorphologists argued over the details of how climate variability might affect streams, as well as the importance of land use versus climate in triggering stream incision and filling. All sides assumed that some change outside the stream system triggered the dramatic responses of cutting and filling.

In the early 1970s a new complication was added to the ongoing arroyo debate. Stanley Schumm and several colleagues published a series of papers demonstrating that the high rates of sediment production in drylands cause a gradual steepening of valley-bottom slopes over time. The small streams that flow down the valley only briefly after a rain are not capable of moving all the sediment introduced from the adjacent hillsides. When the valley-bottom slope reaches some critical threshold, the stream rapidly incises and flushes sediment downstream until a new balance is reached, and the channels once more begin to fill with sediment. This internal threshold can exist regardless of external changes, but it can also be influenced by external changes. In this model of how dryland streams operate, grazing may have exacerbated the turn-of-the-century episode of stream incision but may not necessarily have caused it.

The many potential explanations for arroyo formation provide a striking example of one of the great challenges of geomorphology: equifinality, in which similar results may be generated by very different causes. During the period after World War II, partly in response to the challenge of equifinality, geomorphic research shifted away from an emphasis on cause and effect and toward an emphasis on process. Regardless of what triggered arroyo formation, let's focus on how arroyos form and then refill.

Detouring around the problem of cause and effect may work for a while in basic research, but it presents problems for applied research and management of natural resources. Managers faced with having to make decisions about how many cows to graze per acre or how many acres of trees to cut want to know whether grazing or timber cutting will accelerate erosion and, if so, by how much. Answers are not straightforward. Stocking many cows will certainly reduce vegetation cover and compact the surface. But a lower grazing density may have less effect on hillslope and channel stability than does precipitation variability, or wildfires, or insect infestations, or invasive exotic plants.

Invasive exotic plants were the prime suspect in a research project that brought me back to Arizona nearly fifteen years after I moved to Colorado. Chinle Wash in Canyon de Chelly National Monument runs through a high desert of scrubby brush and grasses widely spaced among patches of bare soil and small sand dunes. The channel is cut slightly below the surroundings, and much of the time it is dry. But to follow the channel upstream is to encounter a surprising change in the landscape. As the Colorado Plateau has been gradually uplifted during the past few million years, the massive, flat-lying sandstone units that are so characteristic of the plateau have in places been warped into gentle folds. Chinle Wash has cut a deep canyon perpendicularly through one of these upwarps, so that as you follow the wash upstream to and into the canyon, sheer walls of orangish red rock rise progressively higher until the sky becomes a narrow swath of blue. During summer, temperatures in the canyon rise above one hundred degrees Fahrenheit, and tall cottonwood trees provide welcome shade and a vivid contrast of green leaves against red rock walls. The cottonwood leaves are equally beautiful when the cooling temperatures of autumn bring out their golden hues and the fallen leaves rustle dryly in the breeze. When winter snow dusts the canyon, it highlights the deposits of sand dunes that now form the rock walls.

The invasive riverside trees tamarisk (*Tamarix* spp.) and Russian olive (*Eleagnathus angustifolia*) were planted in the monument during the 1930s to control erosion. The monument is unusual in that it includes contemporary inhabitants—in this case, Navajo who grow crops and graze cows, sheep, goats, and horses in the canyon bottom. Archeological sites such as cliff dwellings set into alcoves in the vertical canyon walls indicate that people have been living in Canyon de Chelly for thousands of years. Historical photographs from the 1890s to the 1930s show shallow stream channels flowing across the sandy canyon bottom, where only isolated groves of cottonwood and willow broke the broad expanses of sand. Tamarisk and Russian olive were introduced by the Soil Conservation Service and others as a means of limiting the stream's erosion of the low terraces along the valley margins, which provided home sites and fields for the Navajo residents of the canyon. The introduced plants are more tolerant of drought and saline water than the native plants, and they produce more seeds for the winds and birds, such as robins, to disperse. Within a couple of decades, the exotic plants were rapidly spreading up- and downstream along the channels, forming dense thickets that effectively shut out the long vistas of the canyon bottom and increased the resistance of the stream banks to erosion.

By the start of the twenty-first century, tamarisk and Russian olive had covered large segments of the canyon bottom. Streams restricted to narrow channels formed vertical-walled arroyos. Navajo farmers found it difficult to pump the water up to their crops on the adjacent terraces or to get their livestock to water. Moreover, the appearance of the landscape down in the canyon had changed in a manner disturbing to both Navajo residents and park service officials seeking to preserve historical conditions. In many places, as you stood on the banks of the stream, you had to look far up to see the canyon walls above the dense thicket of twenty- to thirty-foot trees that surrounded you.

Snow highlights the crossbeds in the sandstone walls at Canyon
de Chelly, Arizona

The park service assumed that the introduced plants had so
effectively increased the strength of the stream banks that this
had caused the formation of arroyos throughout the monument,
and they undertook an aggressive program of vegetation removal.
They also asked my colleague David Cooper and me to examine the history of channel and vegetation change and to monitor
the channel's response to vegetation removal. The project is still
ongoing, but one thing is clear: the tamarisk and Russian olive,
dense and pervasive as they are, did not cause channel incision.

With graduate student Kris Jaeger, I have spent hours gently
scraping away at the face of a stream terrace in order to collect
charcoal for radiocarbon dating. Like other sites in the western
United States, stream terraces preserved along the margins of the

A view of Canyon de Chelly from the rim, at Spider Rock. Note how the stream in the foreground has no trees growing immediately adjacent to the channel; this part of the valley bottom has been cleared of tamarisk, still visible along the channel in the middle of the photograph.

valley bottom at Canyon de Chelly record recurring episodes during the past ten thousand years in which streams cut arroyos and then refilled them. This indicates the potential for abrupt channel change as internal thresholds are crossed or as regional climate changes. Even more telling, the recent episode of channel cutting at Canyon de Chelly occurred within the time span of repeat aerial photography. When graduate student Dan Cadol examined aerial photographs of the canyon taken between 1935 and the present, he found that some segments of the stream channels in the monument began to narrow and cut down before the introduced vegetation reached those places. Tamarisk and Russian olive may have exacerbated the recent changes in stream channels at Canyon de Chelly, but they did not necessarily cause them. Consequently, removing these trees will not necessarily allow the streams to reclaim their historical appearance as broad, shallow, rapidly shifting channels.

Any drainage basin has complex processes influencing slopes and stream channels. To isolate the role of any single factor, such as grazing or timber cutting, requires either field experiments in which all other factors are held constant while the selected factor varies, or mathematical models in which a single factor can be hypothetically manipulated. Field experiments are challenging because it is hard to hold weather or other factors constant over more than a very short period, and because results tend to be site-specific. Models may lack sufficient details to predict with accuracy the effects of a slight change in a variable such as grazing intensity. Definitive answers are few and far between, with many studies leading to the conclusion that more study is necessary. This frustrates managers, land users, and scientists, particularly within the legal context that drives many land use decisions.

Everyone who cares about the landscape of the West at some point runs up against equifinality. Scientists cannot provide the certainty that resource managers would like. In the absence of that certainty, each individual acts from his or her best understanding. What I find particularly challenging is being aware of the biases inherent in my own understanding. Had I been investigating arroyos during the early decades of the twentieth century, I probably would have worked from the assumption that overgrazing had triggered the widespread channel incision. I might have collected very precise data that quantitatively demonstrated the associations between land use and stream incision, but I might then have missed the key point that those data did not definitively demonstrate that overgrazing causes arroyo formation.

The scientific method's emphases on objective testing of ideas and on reproducibility of results provide a means of attaining knowledge unique among scholarly disciplines. Sometimes this uniqueness is equated with truth. Philosophers question whether truth exists apart from human perceptions. Regardless of the existence of truth, equifinality reminds me that science usually provides imperfect answers.

What Is Natural?

The pine beetle has reached Colorado. I had heard predictions of widespread beetle infestation and tree die-off for the mountainous portion of the state, but the problem seemed unlikely to affect me in the foreseeable future. My attitude quickly changed in 2006, when my annual summer visit to the Fraser Experimental Forest west of the Continental Divide revealed extensive stands of brownish orange trees. I was amazed at the enormity and speed of the change: trees that had appeared healthy the year before were clearly dying or dead. I immediately began to contemplate the implications for the places in Colorado I study. I have been monitoring the distribution of wood in several streams in the Colorado Rockies, including East Saint Louis Creek in Fraser, since 1996. Was I now about to see a huge increase in wood loading within the streams as windstorms knocked over standing dead trees? Would that be followed by a decline in wood within the streams as decades passed before new trees grew to maturity?

Variability over years and over centuries is inherent in natural systems. While I thought about possible outcomes of the beetle outbreak, I read an account of forest ecology in New England. Insect and plant parts fossilized in sediment indicate that about forty-eight hundred years ago the eastern hemlock looper, a native insect, underwent the type of population explosion currently occurring among pine beetles in the western United States. The looper infestation of forty-eight hundred years ago appears to have a triggered a major decrease in eastern hemlock across the tree's range. Although hemlock did not disappear entirely, nearly a thousand years passed before the species started to recover to preoutbreak levels. Meanwhile, sugar maple and beech increased in northern New England and oak in portions of the forest further south. Hemlock did not return as the dominant tree of the New England forests until about fifteen hundred years after the initial decline.

The pine beetle, like the eastern hemlock looper, is native to North America. Some scientists argue that recent fire suppression has created even-aged forests more susceptible to beetle infestations, and that global warming has reduced the severely cold winters that can limit the spread of beetles, but fossil evidence records past population explosions among the beetles and associated tree die-offs.

Episodic insect infestations, variations in precipitation and temperature over decades or centuries, wildfires, landslides, and avalanches each disrupt a forest, and these disruptions influence the rivers that flow through the forest. Ecologists have abandoned the once-popular idea that plant and animal communities attain a stable, climax state in the absence of humans, and geomorphologists have stopped thinking of seemingly stable rivers as being unchanging.

Natural variability creates a challenge for anyone attempting to restore natural systems altered by human activities. If you think in terms of natural systems as opposed to those altered by humans, you have to be able to define what is natural. This can be extraordinarily difficult when forest and stream ecosystems vary over time spans of centuries, apart from any human influence. Fossil or sedimentary records of past variability are usually incomplete, and humans have been in North America for at least twelve thousand years. This suggests that contemporary scientists cannot really recognize natural systems without human influence. Like many other North Americans, I tend to equate natural with good, but what is natural?

No river in the United States is not somehow managed by people. By "managed" I mean that human decisions govern the condition of the river. Management does not imply skillful or effective human control or even deliberate control. If the inhabitants of a region decide not to control vehicle emissions, for example, that decision influences water quality in the region's rivers. Airborne nitrogen from the emissions increases nutrient loading in lakes and streams, altering the biological productivity of these water bodies, as well as the species composition and energy cycling within aquatic communities.

As I write this essay, there is a growing practice in the United States of restoring or rehabilitating rivers. The two words are often used synonymously, although they can be formally differentiated. *Restoration* means returning a river to a condition closely approximate to that prior to some disturbance, whether the recent construction of a highway or the initial settling of the area by people of European descent. *Rehabilitation* often means improving the appearance of a river or putting the river back into good condition. What restoration and rehabilitation entail depends on the prevailing perception about the natural condition of the river, and about what constitutes good condition. Perceptions form the often-unacknowledged basis on which each of us acts. When people with differing perceptions clash over river management, each must confront a personal perception and the biases it imposes.

Contemporary American perceptions of rivers are undoubtedly very different from those of European immigrants two or three centuries ago. Anywhere from six to forty beavers lived along each half mile of river in forested portions of North America before Europeans arrived. Those beavers brought a lot of wood into the rivers and built a lot of dams. Wood ended up in rivers by other means, too. In the uncut forests of North America, enormous logjams developed as trees naturally fell into rivers, from the southeastern coastal plains to the Pacific Northwest. The Red River of Louisiana was famous for its 160-mile-long logjam known as the Great Raft. The first Europeans to reach an area immediately began "snagging" rivers to improve navigation. Most Americans, if faced with these historically wood-rich channels, would now consider them unhealthy oddities.

A group of geographers conducting an ongoing survey among students in university classes around the world show each student the same set of photographs of stream channels. Almost without exception, the students rate the channels with more wood as being more unsightly, dangerous, and in need of restoration than channels containing little wood. Students in Oregon formed the only exception to this trend during the first phase of the study. The geographers conducting the study hypothesized that people living in regions that still had large areas of forest cover might be more aware of the "naturalness" of wood in rivers. The group

Accumulation of wood along Loch Vale Creek in Rocky Mountain National Park, Colorado

asked me to survey students in Colorado next. The Colorado students followed the general trend, rating rivers with wood as less desirable. I now wonder if the difference between Oregon and Colorado reflects the widespread awareness in Oregon of salmon, and the related awareness of the importance of various components of fish habitat, including wood in streams. Colorado does not have any equivalent symbol for the health of streams and watersheds.

While the students in Oregon consider streams loaded with wood to be healthy, government agencies in some states recommend wood removal from streams. Equivalent agencies in other states seek to reintroduce wood to streams in order to improve fish habitat. River beauty is in the eye of the beholder.

These issues were brought home to me when I worked with my graduate student Chris Jaquette on the North Fork Gunnison River in western Colorado. The North Fork exemplifies many of the rivers on the western slope of the Rockies in Colorado. Peaks in the river's headwaters rise above the timberline, and the rocky summits remain snow-covered through the winter. Groves of aspen stand white-trunked among the dark conifer forests, and at the higher elevations large tracts of forest stand between the few thin lines of road that allow skiers to access the headwaters. Lower on the steep flanks of the mountains, precipitation decreases abruptly and the forests of spruce and fir give way to pine woodlands and then to pinyon and juniper widely spaced among shrubs and grasses. Readily weathered and eroded sedimentary rocks underlie all these vegetation communities, and streams such as Muddy Creek are aptly named for the heavy load of silt and sand they carry downstream each year during snowmelt and summer thunderstorms. Farming communities cluster where the larger rivers leave the mountains and spread across the arid flatlands between mountain ranges. The climate and soils here are ideal for growing the sweet, juicy peaches that I love to buy at the farmers' market in Fort Collins.

Our work in the basin was funded by the North Fork River Improvement Association (NFRIA), a citizens' group with funding from the Colorado Water Conservation Board. The group has laudable goals. It wants to rehabilitate the North Fork after nearly two centuries of human alterations. A large coal mine has operated along the upper North Fork for decades. A paved road runs most of the river's length. Farmers have long withdrawn irrigation water from the river and cleared the floodplain forest for cropland. Many of the irrigation diversions are rebuilt each year by running backhoes and bulldozers in the river, reconfiguring the channel so as to direct the flow toward the crops. More than a dozen active gravel mines are spread along several miles of the river. Two towns line the riverbanks.

The North Fork looks messy. Stream banks are eroding in places. The flow runs muddy during snowmelt. The river moves back and forth from side to side across the floodplain in an unpredictable manner, creating many smaller channels that share flow, rather than a single dominant channel. There are few good pools or clean riffles for trout habitat. Members of the NFRIA believe that the river is falling apart, so they set out to stabilize the riverbanks and build structures in the channel that would encourage pools to form. They envisioned a meandering, single-channel river with deep pools separated by riffles on which trout could spawn.

Ecologist David Cooper and I convinced the NFRIA that they should study the history of the river and evaluate the factors that might be controlling its shape, before undertaking any more rehabilitation measures. The group agreed, and funded Chris and another student to study the North Fork's history and geology. What Chris found surprised many people. Certainly the irrigation activities, tree clearing, and gravel mines have not helped to stabilize the North Fork Gunnison, but most signs indicate that it has been a braided river with several shallow, rapidly shifting channels since long before Europeans settled along it.

The North Fork Gunnison drains a region of crumbling rocks. The Mancos Shale is so readily eroded that it almost melts in place, creating huge landslides into the river valley. From forested uplands, tributaries flow the color of café au lait. All of this sediment coming into the North Fork, combined with a flow that swells from a trickle in late winter to a surge during spring snow-melt, has made the North Fork Gunnison a historically unstable river. Under such conditions, the NFRIA is unlikely to maintain a meandering trout stream.

Such streams are in fashion now, and river restorationists go to great lengths to create the ideal meandering channel. Commonly the process starts with backhoes and bulldozers that completely recontour the river channel into a pleasing sinuous form. This form is then fixed in place with riprap and large logs buried in the banks, or perhaps cottonwood or willow seedlings planted in neat rows along the banks. The problem with this design is that meandering streams are inherently mobile. If nature abhors a vacuum, a meandering stream abhors a fixed point. Meanders are constantly changing as the stream erodes the outside of each bend and deposits sediment on the inside of each bend. Individual meanders become ever-more sinuous and then are abruptly cut off at the neck, creating an oxbow lake.

The other problem with creating a meandering stream where one does not exist is that very specific conditions must exist for meanders to persist. If these specific conditions of downstream channel slope, flow levels, sediment movement, and bank stability do not exist, the meanders will "blow out" during the first real flood, and the channel will become braided or straight. Many formerly meandering rivers in Colorado are now braided because upstream mining, roads, and urbanization have changed the amount of water and sediment entering them. To attempt to create a meandering form without addressing these upstream controls is like trying to put a bandage on a hemorrhage without constricting the source of the bleeding.

People actively managing rivers always have some goal in

mind, whether the goal be an agricultural Eden, an esthetically pleasing fishing spot, or a wilderness. As I watch consulting firms and government agencies restore rivers, I think they do not restore so much as manage them to reflect new values. In many cases, the new value represents a version of wilderness thought to exist before contact by industrial-age humans. I have been asked in more than one instance to use geologic records to determine what a river was like before people of European descent altered its appearance. Besides ignoring the impacts that Native Americans or prehistoric peoples might have had on the river—and these impacts in some cases were substantial—this attitude assumes that the river existed in some ideal state to which humans can now return the river. Unfortunately, there is no stationary ideal to sleuth out. Rivers are not necessarily ever in balance, if *balance* implies something static.

Where summer thunderstorms produce flash floods, a river can repeatedly change from a broad, shallow channel with little riverside vegetation following a flood, to a progressively narrower, deeper, and more vegetated channel as time passes. Each new flood resets the river form at unpredictable time intervals. And the entire river system may still be responding to changes in stream flow associated with the steady warming and drying of climate that has occurred since the glaciers retreated ten thousand years ago. Some parts of a river system take a very long time to fully react to a major change such as climatic fluctuation.

Along the North Fork Gunnison River, sixty years' worth of aerial photographs and a much longer sedimentary record suggest that the river becomes broader and more braided during periods of wet climate and high stream flow, and then becomes narrower and more sinuous during dry periods. Human actions can exacerbate or diminish these tendencies, but the tendencies exist regardless of humans.

Those who wish to restore rivers cannot remove all the cities from the Mississippi River drainage or all the dams from the Columbia or the Colorado. We cannot re-create a world that may have existed five hundred years ago. What we can do is seek to restore river function within the constraints that exist in each river's drainage basin. We can let rivers scour their own channels as water and sediment move downstream, and we can plant riverside trees and provide fish with suitable habitats and passage along the river corridor. This type of restoration is much more difficult than building structures along the streambed or banks, because it requires patience, and room for the river to move and change. It also depends on recognition that factors beyond the river channel itself—roads built on hillslopes or houses built on floodplains—will ultimately affect the ability of the river to function. It is politically easier and more immediately satisfying to take a bulldozer into a river and build a channel in two or three months, although it may be very unsatisfying to see the form of this precisely designed channel change back two years later.

As I study rivers across the western United States, I have come to realize that in many cases I cannot specify the natural range of variability that existed before the intensive human use of resources in the West. I cannot quantify what is natural, let alone fully restore it. In the absence of that specific knowledge, I come back to proposing river management that emphasizes allowing rivers to function as freely as possible within the context of continually changing weather patterns, insect infestations, wildfires, and the presence of human communities. Restoration thus involves removing or limiting the constraints on river processes of flooding, channel adjustment, and wood redistribution. We recommended that the NFRIA allow riverside vegetation to regrow and stabilize the banks, and that they limit the use of bulldozers and the construction of instream structures. The North Fork Gunnison deserves a chance to reclaim its own naturalness.

The Disillusioned Angler

The very sitting by the Rivers side, is not only the fittest place for,
but will invite Anglers to Contemplation.
Izaak Walton, *The Compleat Angler*

The river smells like life this autumn morning. From where I
stand, on a cobble bar inside a bend, scents drift past me, min-
gling delicately. Cottonwood and willow leaves are falling. An
earthy scent of leaves returning to the soil rises from the river's
edge. The river level too is falling, exposing algae that dries into
a musty-smelling, hardened brown crust on the rocks. When I
slip on the rocks and expose the moist underlayer, it smells like
bruised lettuce. The sun's warmth still calls up pinesap, and that
aromatic scent drifts across the river, too. The water itself smells
faintly of fish and insects—just the lightest of odors, a hint of the
life hidden in the shadows of pools.

A grove of tall, golden-leafed cottonwoods lines the river channel, a reminder of the big flood fifty years ago that deposited the
sediment on which they germinated. For half a century now the
trees have remained undisturbed by floods. Their stately presence

The Cache la Poudre River in autumn

gives the appearance of permanence. Beyond the cottonwoods rise dark-green slopes covered with pine, spruce, fir, and the occasional chute of aspens flowing bronze down the valley side. The sky is a satisfying blue above the river valley.

The sun is noticeably farther south each day. Sunlight shatters into swiftly changing mosaics among the trees. The river flows quietly, touched green and brown by undulating sunlight and shadow. Tiny ripple crests catch the light and toss it back in diamonds that glimmer for an instant. Where the river descends a steeper rapids downstream, water droplets burst upward and momentarily free themselves from the river like salmon leaping a falls. And then, miraculous as it always seems, a trout leaps from the water to catch an insect and the river becomes truly alive. The slender, silver grace of the fish's leap appears only at the edge of vision, but spreading ripples bear witness.

I walk downstream toward the spot where the trout jumped, enjoying my solitude yet drawn to this sign of life. Before I am halfway there, a fisherman walking upstream comes into view. His appearance matches what I expect of a fly fisherman: he wears a handsome leather fedora, chest-high waders snugged around his waist, and a hoop net hanging down over his fishing vest. His khaki and brown clothing blends well with the river and the morning light. He carries a long, supple rod carefully in his hands. The slender fishing line forms a graceful loop beneath the rod.

He stops near the spot where the trout jumped and begins to play out line in overhead casts. His body moves with an economy of motion, as though the rod and line were an extension of his arm and shoulder. He places the fly lightly at the upstream edge of a deep run. The fly floats downstream as fitfully as a real insect, but he lifts and flicks it repeatedly low over the water before placing it delicately once more. Such enticement the trout cannot resist. On the fourth cast the fish rises and strikes the fly.

Now the dance enters a new phase. The fisherman lets the fish twist and surge through the water, playing out line as needed, but he maintains a steady pressure that drags the fish relentlessly back. After several minutes, he cradles the exhausted fish with his net in the shallows. Admiring the rainbow colors splashed along the trout's flanks, the fisherman releases it back into the river to swim free once more. The man straightens, replaces his net, and walks downstream to seek more fish.

I know the pleasure of the fisherman's morning. I too used to fly-fish, many years ago. To catch a fish is to share in a river's secrets and pull it into your own world of light and air. This life largely hidden from view as you stand by the bank comes forth because of your skill in deception. The artificial fly that you have created, chosen, and deftly manipulated has attracted this fish with millions of years of instinct behind it. The adrenaline and excitement in the contest of actually landing the fish give way to the satisfaction of food, or a trophy.

I no longer fish. My reasons go against the well-established traditions of fishing and the literary genre they have given rise to. Anglers, particularly those who fly-fish, endow their pastime with a quasi-religious mystique. But this is not my religion. I have learned too much about its inner workings to graciously accept it any longer. I do not fish for three reasons. First, I do not like to kill fish, and hooking and then releasing the fish is likely to kill it anyway. Second, I do not like to eat fish. Third, and most important, I do not like the machinations necessary to enable me to fish.

Like Virginia Woolf's three guineas, these three reasons have long explanations attached to them. The first two reasons are easiest to explain. I do not like to kill fish because I do not enjoy taking life. Only 2 to 5 percent of the fish caught with a fly and then released die from the experience, although fish in heavily used waters may be caught repeatedly in a single day. Fish caught with bait are more likely to swallow the hook and, on average, half of these fish die once they are released.

I eat meat and fish, so I consume creatures whose lives have been taken by other people. But I take no pleasure in actually ending another creature's life. A fish is beautiful and exciting so long as it lives and swims in the river. Once it is pulled ashore, its struggle to breathe and its thrashing hold no beauty. The shining colors fade as the animal's life ebbs and the vibrant energy is stilled.

We do not quite have "combat fishing" in Colorado, where eager, aggressive anglers line the stream banks almost shoulder to shoulder to create a gauntlet that the fish must run. I have seen this kind of fishing in Alaska when the salmon spawn, as well as the accompanying trash, trampled banks, and impromptu roads. But we have something close enough here in the streams of the southern Rockies.

The Cache la Poudre River flows clear and cold down a mountain canyon with easy road access. On any weekend the roadside is lined with the cars of anglers. They fish for *Oncorhynchus mykiss,* the beautiful rainbow trout native to the portions of western North America bordering the Pacific Ocean. The rainbow are not native to the Poudre, and neither are *Salmo trutta,* brown trout from Europe and Asia, and *Salvelinus fontinalis,* brook trout from eastern North America, which are also present along the river. *Oncorhynchus clarki,* the greenback cutthroat trout native to this river, is now a threatened subspecies restricted to headwaters where the nonnatives have not been stocked or cannot reach because of steep rapids or waterfalls.

All the nonnative species were introduced to Colorado during the late 1800s to improve the fishing. The greenback cutthroats had been fished out or displaced by the devastating sediment loads set loose by placer mining. Anglers prefer rainbow trout because these aggressive fish fight the hook more forcefully. These trout grow rapidly and were able to out-compete the remaining natives. And fish biologists knew more about raising rainbow, brook, and brown in hatcheries. Vigorous stocking programs continue today. As of 2003, the Colorado Division of Wildlife was annually dumping seventy thousand rainbows into thirty miles of river in upper Poudre Canyon, as well as another six thousand fish into the river in the lower canyon, where many of the campgrounds are located. In addition to preventing reestablishment of the native trout, this and other stocking programs have spread infections, such as whirling disease, throughout fish populations.

Early one morning in August 2004, I visited one of my long-term monitoring sites in Rocky Mountain National Park. Once a year I climb to this site at ninety-seven hundred feet along the headwaters of Ouzel Creek to check on the pieces of wood in the creek that I first surveyed in 1996. I am curious about how frequently the wood moves and how long each piece will remain within the two-hundred-foot-long study reach.

Although it is only August, summer is short in the high country, and I find that gold already tinges the aspen leaves on this clear and cool morning. Descending to the creek, which cannot be seen from the trail, I am surprised to find several biologists from the U.S. Fish and Wildlife Service working at the normally deserted site. They are electrofishing, moving a small electrified wand systematically back and forth as they wade upstream, in order to stun the fish sufficiently that they float to the surface and can be counted. As I walk up, I see that the biologists have a ten-gallon bucket full of trout less than a foot in length. To my surprise, they are all brook trout. This segment of Ouzel Creek is immediately upstream from a twenty-foot waterfall, and I had not expected to see any exotic fish here. The brown and rainbow trout are mostly under control and have been eliminated from this creek, but the aggressive brook trout remain so numerous that many are in poor condition, as the biologists demonstrate by pulling out skinny trout with oversized heads.

Ouzel Creek, like others in the park, was poisoned with anti-mycin in the past to kill the nonnative fish. Some of the brook trout survived, however, and now they are taking over streams. When I ask why the streams can't be poisoned again, the biologists reply that anglers pressure the park service to keep the "fighting brookies." Without such pressure, the brook trout could be eliminated. This prompts the obvious question of why angling is allowed in national parks where hunting is banned. One fish biologist shrugs, then tells me that a few days earlier two mountain goats wandered into the park from the herd living on nearby Mount Evans. The goats, which were released on the Forest Service's Mount Evans decades ago, are not native to this part of the Rockies. The two strays were promptly tracked and shot by park service personnel.

The contemporary double standard for managing furry and finny species in the park reflects a long history of conflicting attitudes toward native fish. Greenback cutthroat, the only native trout in the park, provided a rich food resource for the first European settlers in the region. The Estes family, for whom the town of Estes Park along the national park's eastern border is named, caught thousands of fish that they took to markets in Denver. They even fished through the ice in winter. In 1875 Horace Ferguson caught 270 cutthroats from a single beaver pond over the course of three days. This kind of fishing was too good to remain a secret, and unregulated recreational fishing expanded dramatically starting in the 1870s. Contemporary newspaper accounts indicate that many of these anglers would catch up to a hundred fish in a day and keep all of them.

The first state fish hatchery was built near Denver in 1881. Fishery biologists imported a variety of nonnative species, searching for the fish with the best hatching rate that would grow quickly when stocked in the wild. Brook trout met the criteria. In 1886, twenty thousand brook trout were released into the North Saint Vrain Creek drainage, which includes Ouzel Creek. Despite subsequent additional stockings, fish populations became depleted by 1895. The male citizens of Estes Park formed a trout protective association in response, with the intent of building a hatchery. Nothing much actually happened until their wives took over the fund-raising for the operation. The women were so successful in raising money that the private Fall River fish hatchery opened nearby in 1907. The Fall River hatchery increased stocking rates dramatically by releasing a million trout each year. Rainbow and brook trout, as well as cutthroats from Yellowstone National Park, were released. Enthusiasm for stocking grew so high that local residents packed fish in to remote, high-elevation lakes that had been without any fish.

Swamped by these huge numbers of exotic species, the greenbacks were thought to be extinct in the park by the 1930s. The hatchery operated until 1982, and the intensive stocking of nonnative fish by private and government interests continued within the park until 1968. Management shifted in favor of native greenbacks in the early 1970s, but as my experience at Ouzel Creek indicates, the natives cling by a "fin-hold." When I asked the biologists if they were going to kill the brook trout they had just laboriously removed from Ouzel Creek, they shook their heads quickly. That was not part of their task for the day.

Overfishing, competition from introduced species, and habitat loss make it very difficult for cutthroat trout to recolonize their former range. Cutthroat, designated the Colorado state fish, now occupy less than 5 percent of their historical native range. At intermediate elevations in the Rockies, much of this range includes sections of private land where landowners are reluctant to admit the existence of a threatened or endangered species for fear this would restrict their land use options.

As consumptive outdoor recreation goes, fishing is very popular in Colorado, but it is not a big revenue generator. State residents pay twenty dollars for an annual license, and nonresidents pay forty dollars. The fishing program is subsidized annually to the tune of five or six million dollars by elk hunting, the big moneymaker in Colorado. But anglers seem to have a disproportionate level of clout in the politics of outdoor recreation and resource management. Trout Unlimited and other sportfishers' groups are not shy about making their wishes known. Trout Unlimited can be a force for stream protection when speaking out against the deleterious effects of cattle grazing in the riparian zone, for example. The organization helps to educate people about the importance of healthy, functional stream ecosystems, and it funded some of my research on stream recovery along the North Fork of the Cache la Poudre River following a reservoir sediment release. But this group and other fishing lobbies sometimes lose sight of the stream as an ecosystem and focus solely on the sport of fishing. The anglers who catch trophy trout in the unnaturally clear, cold waters released from the base of Glen Canyon Dam on the Colorado River lobby hard to prevent any change in dam operation that would re-create the historically warm, turbid flows necessary to the survival of endangered native fish species on this stretch of the river.

North America has about 800 native species of freshwater fish. Six hundred species are present east of the Continental Divide, and about 40 species are found on both sides of the Divide, leaving only about 170 species native to western North America alone. Many of these species exist only in a single small spring or within a single river drainage basin. The West is poorer in fish fauna than the eastern portion of the continent because of the West's long history of disruptive geologic and climatic events that reduced the diversity, availability, and reliability of river and lake habitats. And now the West's fish diversity is being lowered drastically by extinctions. Nearly half of North America's more than 200 species of freshwater fish listed as threatened or endangered live in the West, as did 17 of the 30 species already extinct. Of these, Colorado has lost the yellowfin cutthroat trout *(Oncorhynchus clarki macdonaldi),* the extinct cousin of the threatened greenback cutthroat.

No one knows how many nonnative fish have been introduced to western waters for sportfishing or other purposes. Many of the introductions occurred before fishery biologists even described the native fish fauna. Too many people have believed and continue to believe that they perform a public, or at least personal, service by releasing fish as they go, like piscivorous Johnny Appleseeds. These fish, freed without much forethought or understanding of ecosystem dynamics, have played havoc with streams throughout the country. The introduced fish may have no effect on native species, but more often they result in reduced growth and survival, or even elimination, of the natives. Many studies document changes in native fish communities occurring with the introduction of other species, but biologists often have a difficult time separating the effects of nonnatives from the simultaneous effects of habitat loss and overfishing.

The biological effects of stocking nonnative species are only part of the story. Native fish must attempt to compete in streams engineered to create habitat favorable to game fish. Many a braided mountain river has been bulldozed, riprapped, and strewn with artificially placed boulders and logs to create pools for trout habitat. Anglers tend to believe that braided rivers provide poor trout habitat, despite their wealth of shallow backwaters that serve as nurseries for young fish and retreats for older fish. When avid anglers see a braided river, they too often see a river that seems to need help in creating pools. And because they also tend to see river *form,* rather than function, they see a river that can be "improved" with a fluvial straitjacket of boulders and concrete. Ironically, such a straitjacket is likely to ultimately reduce the river's ability to create pools. My former student Doug Thompson visited sites along Connecticut's Blackledge River that were altered by the Civilian Conservation Corps during the 1930s to "improve" fish habitat. Sixty years later, these sites had poorer habitat than portions of the river that had not been reconfigured.

An extreme example of stream engineering for fish production comes from contemporary rehabilitation projects on the Little Snake River in northwestern Colorado. Here, on private lands, consultants have installed a series of vertical drops along the streambed using rocks and logs. This portion of the Little Snake flows along a broad valley that is too gentle to support naturally formed steps and pools. Artificial structures, along with feeding of the resident fish, have produced a sort of übertrout disproportionately large relative to the size of the stream, which anglers pay a lot of money to catch.

When fishing is taken to extremes, the fish become trophies, not living things. And in the eyes of some people, anything that will help them win trophies becomes justifiable—from extirpating native species to engineering streams as though they were fish-producing factories.

I do not fish, because I respect streams. I do not need a fishing rod to invite contemplation. The moving water alone suffices. The sight of a fish in the water, its fins undulating rhythmically from its sleek body, is a gift. I need not ask for more.

Poisoning the Well

At the heart of Colorado lies an enormous water tank known as the Rocky Mountains. Winter snowfall on the mountains reaches depths unimaginable on either the plains to the east or the plateaus to the west. As the hours of daylight grow longer and the air temperatures warmer, this snow melts and flows down hundreds of rivers to the lowlands lying on all sides of the mountains. Clean water from the mountains is the lifeblood of these lower regions, yet many of the land uses in Colorado have the effect of substantially reducing the flow and contaminating the water.

So long as water flows from the mountains, all is well. But if the snows do not fall in sufficient quantity and the water does not flow, every living creature suffers. Mountain soils grow parched. Plants die or go dormant, failing to flower and set seed. Litter on the forest floor becomes tinder dry, and the moisture stored in the trees declines. Forest fires thicken the hot summer days with smoke. Springs dry up and the little creeks flow for a shorter time or do not flow at all. Mountain animals that eat the seeds and fruits starve, and without running springs and creeks, they dehydrate. Birds grow disoriented and fall from the sky, panting, to die within minutes. Some of the starving animals travel to elevations lower than the ones they normally visit, desperately seeking sustenance. Elk come to graze on suburban lawns. Black bears visit the garbage cans and fruit trees of towns and cities below the mountains. When the bears are shot, wildlife officers find that the animals have less than a quarter of their normal body fat.

Lower elevations suffer, too. The spring and summer high flows from the mountains that set the pace of life for plants and animals along the lowland streams do not come. The plants go dormant and do not reproduce, or they die. Fish living in rivers that have already been reduced to a trickle by our withdrawals of water end up marooned as streams are sucked completely dry. What little flow remains in the rivers is quickly taken by thirsty human communities, but farmers still lose crops and suburban lawns go brown under water rationing. Drought spreads downstream like the ghost of a flood, bringing hardship to all who depend on mountain snowmelt.

I have seen all these changes in Colorado during the past few years. We rely on the enormous water tank of the Rocky Mountains. When it goes dry we start to speak of limiting population growth in the state, of building more reservoirs to store every precious drop of water during drought years, or even of seeding winter storm clouds with silver nitrate to coax more water from the sky as the clouds race past on their way to Nebraska.

I write this in mid-November, having just returned from a day of skiing in the mountains. The snow is already more abundant than it has been during the past few winters, and it continues to fall steadily. I am happier than I have been since the drought began. But this could be a false start. The snow, having begun so promisingly, could stop. Or a heat wave during some part of the winter or spring could melt the snow too soon for the growing season, or too rapidly to recharge the soil. But I remain optimistic. An El Niño circulation system is raising the hopes of many in the Southwest, and I think this will be a good snow year in Colorado.

．　．　．

The flow paths of water from the mountains to the plains epitomize much of what I have come to understand about the connectedness of disparate landscapes such as mountains and plains, as well as the connectedness of humans and landscape. We do not simply live on a landscape. We depend on it and alter it as our needs and expectations change. To trace the journey of a snowflake falling in the Rockies and then downstream to the eastern plains is to explore these connections and gain a new perspective on one's sense of place in the American West.

The journey begins on a hillslope. It might be above the timberline, where the sculpting power of nearly constant wind carves snowdrifts into hard cornices. Or the slope might lie where tall spruce and fir trees disrupt the wind, allowing large, delicate flakes to drift down into thick layers of powder. Regardless, the snowflake will probably remain in place throughout the winter, gradually buried by newer flakes until it settles into the denser, granular base of the snowpack.

The mountain snowpack forms an anxiously watched resource in terms of both quality and quantity. Ski resorts can supplement the natural snow supply with snowmaking machines, but the manufactured snow is a poor substitute for natural powder.

Regional water boards and water conservation districts track the snowpack diligently, basing their forecasts of water availability and operating strategies on its water content. Local newspapers report the evolving snowpack as the year progresses.

In the Colorado Rocky Mountains, the snowpack joined by our snowflake likely rests on a hillslope that is part of a national park or a national forest. The forest has probably been cut more than once, and roads may interrupt the hillslope, bringing people to hunt, camp, or ski in the area. But the surroundings remain largely undeveloped, and any contamination that the snowpack receives comes mainly from the air.

Some of the contaminants take the form of nitrogen blowing up from the farm fields and large cities at the base of the mountains. This nitrogen, which originates from auto exhaust, coal-fired power plants, gas and oil wells, crop fertilizers, and livestock manure, falls onto the mountains in rain and snow. Twenty years of monitoring at the Loch Vale watershed in Rocky Mountain National Park indicates an annual increase of 2 percent in nitrogen deposition. This does not sound like much, but in 2004 scientists warned that a threshold was being approached. The scientists detected changes in soil chemistry and an increase in formerly rare algal blooms in alpine lakes. They also found altered growth in three-hundred- to seven-hundred-year-old Engelmann spruce and excess nitrogen flushing from soils into the streams. Scientists predict that, within a decade, soil acidification will weaken and kill trees and acidify high-altitude streams to the point that fish will die.

Dealing with the problem is politically difficult because of the diverse contributors to excess nitrogen. Vehicle emissions account for 45 percent, and these come mostly from the cities at the east-

Loch Vale, Rocky Mountain National Park

ern base of the Rockies, where population jumped from 1.9 to 2.9 million between 1980 and 2000. Coal-fired power plants produce another 34 percent. Much of the rest comes from agriculture. Irrigated acreage in Colorado has increased 73 percent since the late 1950s, as the South Platte River basin has become a national center for cattle feedlot operations. Cores of sediment from high-altitude lakes in Rocky Mountain National Park indicate that nitrogen levels remained essentially unchanged for fourteen thousand years, and then abruptly increased during the 1950s, when these contributors began to increase.

A study published early in 2008 documented changes that are even more disturbing. Scientists with the Western Airborne Contaminants Assessment Project detected highly toxic mercury, the banned insecticides DDT and dieldrin, and a flame-retardant chemical in the waters of eight national parks in the western United States. The scientists also found intersex trout—fish with both male and female reproductive traits as a result of exposure to endocrine-disrupting chemicals—at five lakes in Rocky Mountain National Park. Fish at all the national parks, even Alaska's Denali, contained mercury at levels above what is considered safe for consumption by birds, and some of the mercury levels exceeded the lower limits for human consumption. Much of the mercury comes from coal-burning power plants, and this mercury can be redistributed by wildfires when mercury absorbed by trees is released in a wildfire's intense heat as gaseous elemental mercury.

.   .   .

With the warming of spring, the snowpack thins, sublimating into water vapor at the surface and melting into liquid water at the base. Water flows downhill in uncountable tiny rivulets and marshy seeps, collecting into small headwater creeks. The creeks may meander across emerald green alpine meadows, where quickly growing flowers spread flushes of vivid color during spring and summer. Or the creeks may fall precipitously down steep slopes over bedrock drops and through bouldery cascades.

Many of the creeks will pause in the limpid blue-and-green water of an alpine lake. These lakes were the collecting point for snow that metamorphosed into glacial ice during the last ice age. At that time, depressions along the headwaters of a valley collected snow that persisted through the summer. Under the pressure of the steadily accumulating snow, the lower layers melted and then recrystalized as ice. Some of the meltwater filtered into cracks in the underlying rock and refroze. The water's expansion while freezing further fractured the rock, creating a layer of angular rubble beneath the ice. With sufficient mass, the ice began to flow down the valley, plucking the rocks below and scooping out a deep bowl that filled with water as the glacier melted.

Some of these alpine lakes hold fish that dart like dark shadows through the numbingly cold water. If the lake is high in elevation and has steep, swift creeks and waterfalls downstream from its outlet, it probably did not have fish before they were stocked by anglers. Like the snowpack, the lake waters are probably contaminated with nitrogen carried up from the lowlands on the winds. The extra nitrogen causes algal blooms in the lake, but the water quality remains fairly good.

Beyond the lakes, the creeks grow larger. They still flow in steep channels, rushing downstream with a whitewater impetuosity that only the ouzels brave. Ouzels *(Cinclus mexicanus)* are river birds, born on rivers and never traveling more than a few hundred yards from water. Superbly adapted to riverine life, ouzels possess an extra supply of red blood cells and thick insulating feathers that allow them to dive for minutes in the icy waters while hunting insects, and special eye muscles that allow them to see equally well above and below water. Bobbing repeatedly beside a river, an ouzel personifies the restlessness of mountain water.

The ouzel may also personify the hidden history of the mountain stream, for the seemingly vibrant bird can be weakened by toxins carried in its body. Miners traveling from the eastern United States to the California goldfields found gold along the rivers at the base of the Colorado Rockies. But they had their sights fixed on California, and kept going. When things didn't pan out in California, they remembered the Colorado find, and returned in 1859. The discovery of placer gold at Cherry Creek near Denver set off a series of rushes in gold, silver, copper, lead, and zinc that lasted well into the twentieth century.

Miners found "pay dirt" along a broad swath of mountain streams, from the Poudre River near the Wyoming border, down to southern Colorado. In order to separate the placer metals from the valley-bottom sediments, the miners sluiced the sediments down long flumes with ridged bottoms that collected the heavier metals. Sediments were scooped across the entire valley bottom, all the way down to the bedrock, and were shoveled into the flumes. The massive destabilization associated with this activity caused large amounts of sediment to travel downstream, turning the clear mountain waters to muddy sluices, filling in pools, and killing fish and aquatic insects. Once the river channel lost its internal cohesion, the river remained unstable for decades after mining ceased.

My student Marsha Hilmes and I studied the Middle Fork of the South Platte, where mining continued into the 1950s. Historical records and unmined river segments upstream indicate that before it was mined the Middle Fork was a deep, narrow stream meandering through broad meadows of willows and grasses. Beaver dams created ponds that helped to stabilize the river, as did the layer of silt and clay tightly bound by grass roots near the top of the riverbanks.

Mining ripped out the vegetation and the cohesive silts and clays, leaving cobbles and boulders that neither trapped fine sediment nor supported riverside vegetation. The Middle Fork became a broad, braided river, constantly shifting back and forth, creating new islands or bars and then destroying them. Aquatic habitat declined dramatically, and so did species density and diversity.

Miners used dredge boats during the final stages of mining on the Middle Fork. A dredge boat operation uses a giant shovel to scoop up the river bottom, processes the metal-sediment mixture, then dumps the waste sediment, leaving a line of mounded sediments behind as the boat moves down the river. One cloudy day as I flew cross-country, the clouds parted as the plane crossed the Rockies. I glanced down to lines of pale, lumpy terrain that looked as though an enormous gopher had been burrowing across the Earth, and instantly knew we were over the Middle Fork.

The nineteenth-century miners also used mercury to extract gold from the valley-bottom sediments. Mercury amalgamates with flecks of gold, making the flecks easier to separate from the sediment. But mercury is also extremely toxic to living organisms. Even a tiny amount can be deadly, and mercury persists in the environment for centuries. River segments contaminated by mining may have recovered their channel forms and riverside vegetation, and their appearance may differ only subtly from that of unmined rivers. But the mercury and other toxic metals such as lead and cadmium associated with mining persist, working their way through the aquatic food web from silts and clays to insects, and from there to ouzels and fish.

Ouzels living along formerly mined streams have in their blood anomalously low concentrations of aminolevulinate, delta-, dehydratase, typically referred to as ALAD, an enzyme that produces a component of the hemoglobin that carries oxygen in red blood cells. Exposure to lead inhibits ALAD, which is required by all cells. When ouzels have less ALAD, and lower hemoglobin levels, it may reduce their ability to make frequent, prolonged dives for food in frigid mountain streams.

The presence of ouzels along a stream is sometimes a sign of a healthy stream ecosystem. The birds cannot survive along severely contaminated streams, because there are insufficient aquatic insects to eat and because the contaminants in insects accumulate in the birds, eventually poisoning them. The detection of ALAD in the blood of ouzels living along formerly mined streams indicates that the birds can live with some level of contamination, but we do not know how this compromises their health and life spans.

. . .

The downstream journey of a mountain creek may be interrupted when the creek is still too small to support fish. Despite the airborne contaminants, waters of the highlands are relatively clean unless they drain an abandoned mining site, but as they flow downstream the waters grow progressively more contaminated with human wastes, and progressively more of the water is diverted by people living along the river corridors. Those who manage the water supply for Denver and other cities at the base of the mountains seek to divert water at ever-higher elevations, siphoning streams at their sources into canals and pipelines that bring the water directly to the cities. Such diversions are illegal in the national parks, for they destroy the river ecosystems that the parks were designated to protect. But the Organic Act authorizing the national forestlands said nothing about ecosystem preservation. Authorization for the original forest reserves focused on protecting timber supplies and watershed functions because, if the forest remained in place, it helped alleviate downstream flooding and sedimentation.

As the water users climbed higher into the mountains in their search for diversion points, the Forest Service sought to stop them. The Service could not argue for preserving instream flows based on biological criteria, for its authorizing legislation did not specify preserving fish communities or riverside vegetation. Instead, the Service decided to argue for leaving water in rivers based on channel maintenance and flood conveyance. Because the forest preserves were intended to protect watershed functions, requests for water could be denied if the Service could demonstrate that the diversions would clog channels with sediment and increase flood hazards.

The Denver Water Board and the U.S. Forest Service went to the Colorado Supreme Court in 1989 to argue over whether river channels needed water. The Denver Water Board maintained that the streams of the Colorado Rockies were largely formed by much higher flows during the melting of ancient glaciers. Contemporary flows did little to shape the channels, and flood hazard would not be increased by diversions that left the downstream channel segment completely dry, especially if an occasional large flow was allowed to flush accumulated sediment. The Forest Service maintained that the mountain streams were actively forming, and that they reflected contemporary flows that continually moved sediment from adjacent hillslopes. The amount of sediment that moved corresponded directly to the volume and duration of flow in the river channels. The Forest Service lost the case.

I had just begun teaching river geomorphology at Colorado State University during the months when the case was being tried. The lead expert witness for each side was a famous river geomorphologist. These two men had largely shaped the discipline during the later twentieth century, and I had learned my profession by absorbing their words and ideas. Each man argued from strong convictions about how rivers worked and about how water in the West should be used. I summarized the ongoing case carefully in my lectures, unwilling to take a definitive stand.

Both sides in the case had relatively few data on which to

East Saint Louis Creek. The line above the stream channel is one of our survey markers.

base their arguments. They extrapolated from studies of rivers in other regions, and they hastily collected data from Colorado's mountain streams. Although the Forest Service lost this case in Colorado, they were prepared to try again in other western states with similar water diversions. They established a more thorough program of studying the geomorphology of mountain streams. I was invited to work at East Saint Louis Creek in the Fraser Experimental Forest of Colorado shortly after the 1989 court decision. I have been working there, along with several generations of graduate students, ever since.

By the time it reaches the green metal gate of the Denver Water Board's diversion structure, East Saint Louis Creek has dropped almost three thousand feet in three miles. The creek has come from cold, windy slopes above the tree line, down into old-growth spruce and fir with moss-hung trunks. The channel is only a few feet wide—narrow enough to jump across if you are agile. But during spring snowmelt, the water flows with such force that in the deepest parts I can barely stand upright while taking measurements. Left to its own devices, the creek would drop another mile down to its junction with Saint Louis Creek. Instead, flow in the creek abruptly ceases when the diversion gate is open.

Our work began with studies of how water levels along the creek above the diversion relate to sediment movement. We surveyed cross sections of the channel and measured current velocity. We tried to trap sand and gravel moving downstream along the creek bed by using little metal funnels with mesh nets behind them. Our sediment traps looked like underwater butterfly nets for sediment, except that we held the "nets" stationary and let the sediment come to us. The work was cold and strenuous. Three graduate students—Kathy Adenlof, Clifford Blizard, and Carolyn Trayler—spent three successive summers in the forest, working in the creek by day and sleeping at a nearby Forest Service dorm. The Forest Service also had its own employees doing measurements in other streams, and together we gradually assembled a story of how the mountain streams work.

The high flows produced as the glaciers melted were capable of moving much larger boulders than most contemporary flows can move. Many large boulders were also released as the glaciers melted, and have subsequently been moved relatively little. But the big boulders that form most of the framework for a mountain stream do move. Every few decades, a particularly large flood or a debris flow destabilizes large sections of the stream channel, and there is a period of reorganization before the channel becomes stable once more.

Wood falling into a creek can also cause dramatic changes, as I and my graduate students Janet Curran and Andrew Wilcox have seen in our studies of wood in streams. When a tree falls across a small mountain creek, the tree trunk is so much longer than the width of the channel that stream flow in the creek is not capable of moving the wood. With time the tree decays and breaks into smaller pieces that the water can carry downstream. These pieces, which can still be several feet long, collect in debris jams that store sediment and form small waterfalls in the channel. When a key piece in a debris jam breaks, a mass of water, sediment, and wood surges down the channel, mimicking the effects of a flood or debris flow.

Between these major disturbances the stream remains fairly stable. Water flowing downstream during spring snowmelt does not move the largest boulders, but it does carry sand and gravel and fist-size cobbles. The bed and banks of the channel are rough and irregular with large boulders and log fragments. These irregularities create a turbulent flow with short bursts of rapid velocity and jets of shooting water next to slowly recirculating eddies. The episodic movement of sediment downstream makes it very difficult to measure or predict. But the longer the spring high flows last, the more sediment moves downstream. The Forest Service has a large collection pond along East Saint Louis Creek, just above the diversion gate. For several decades the Service has measured how much sediment enters the pond each year, and then compared this to the volume of water flowing down the creek. On average, each year the creek pumps out a ton of sediment for every four-tenths of a square mile in its drainage basin. Clearly, water flowing steadily along a mountain stream maintains the capacity of the channel to convey large amounts of water during floods.

Subsequent studies have demonstrated the resilience of these creeks and the organisms living in them. The steepest segments of a creek, where water cascades down over a series of steps formed by logs and very large boulders, show little change in form despite several decades of diversions, particularly if an occasional large flow is permitted along the stream. Gentler segments where the channel flows across pools and riffles are more likely to narrow and store sediment as a result of diversions and decreased flow, suggesting that maintenance of these segments requires that there be less diversion. Perhaps more surprising than the stream's physical resilience is that the aquatic insects can maintain their abundance and diversity fairly well if groundwater naturally seeping into the channel downstream from the diversion augments the stream flow, or if limited high flows are allowed to pass downstream each year.

. . .

Not every mountain creek in the Rockies is as thoroughly stopped in its downstream journey as East Saint Louis Creek. Some of the creeks and rivers continue to flow all the way to the base of the mountains, although the level and timing of flow may be controlled by an elaborate system of plumbing. Much of the water flowing down the eastern slope of the Rockies has been diverted from the headwaters of the Colorado River and the western slope. A spaghetti-tangle of subterranean pipes and aboveground ditches sends water through and over and under the mountains to the farms and cities stretching along the Front Range from Fort Collins south to Colorado Springs.

All this water engineering plays havoc with the natural functions of a creek, making the Front Range a very good place to examine human activities on streams. But the problem with this approach is that, before I add an extra layer of complexity, I would like to know how mountain streams function in the absence of humans. There has simply been too little study of mountain streams to provide that basic level of knowledge. So I continue to search for the relatively pristine portions of each river drainage I study as a baseline for comparison. Some of these little-altered river segments lie within Rocky Mountain National Park. Privately owned water diversions and dams existed as inholdings in the park when it was designated in 1911, but the Park Service has been gradually acquiring and removing these interruptions, returning streams to a more natural state.

North Saint Vrain Creek heads in Rocky Mountain National Park and flows uninterrupted to a reservoir at the base of the mountains. My work along North Saint Vrain began with my graduate student Doug Thompson, who examined how the movement of water and sediment changes between pools and riffles.

Much of the new river restoration being conducted around the country focuses on enhancing the formation of pools, because pools provide critical fish habitat. Consultants use a backhoe to excavate a deep pool every few hundred yards, but if these pools do not mimic the periodic deeps created by stream flows and sediment movement along the river, the artificial pools will erode along the banks and fill with sediment in a few years. Creating sustainable pools that will be maintained by river processes requires studying the river and understanding its processes. Geomorphologists and engineers are not quite at this level of understanding yet, which is part of what motivated Doug to work on the pools and riffles of North Saint Vrain Creek.

Doug became intimate with one particular pool on North Saint Vrain Creek during the four years he spent writing his master's thesis and doctoral dissertation. He used nontoxic resins to coat some of the creek's cobbles with neon colors, to the surprise of anglers who wandered into the area, and then tracked the movement of these cobbles over succeeding spring snowmelt seasons. He waded into the swift, icy waters to measure velocity and photograph sediment movement with underwater cameras. He re-created the pool in an experimental flume at the university and measured changes in water velocity and sediment movement as he systematically changed flow levels and pool configurations. He calibrated a computer model using the field and flume data and then simulated further "indirect" experiments with the model to study the role of pool geometry and flow levels on pool processes. As the product of all this effort, he developed a conceptual model of how pools are formed and maintained.

An obstacle such as a very large boulder constricts the stream flow from the side. This allows slower-moving water to pond up- and downstream from the obstacle in eddies, and it creates a jet of fast-moving water in the center of the channel. The jet of rapid flow is effective at scouring the streambed and creating the depression that becomes a pool. A larger lateral constriction produces a deeper pool. Sediment moving into the pool can become trapped in the marginal eddies, particularly during higher flows, when the boundaries between the central jet and the marginal eddies create zones of intense turbulence. As flow decreases, some of the sediment stored in the eddies can slump back into the center of the pool to be flushed downstream during the next high flow. Coarser sediment such as cobbles and gravel moves episodically downstream during each high flow, traveling from a pool to the next shallow riffle, and then on to the next pool during a subsequent high flow.

With later graduate students, such as Jaime Goode, I used this conceptual model to examine rivers with human impacts and ask questions such as: How does flow diversion alter pools? Are declines in pool size greater along rivers where adjacent roads increase sediment entering the river, or along rivers where nineteenth-century mining produced huge influxes of sediment? Naturally functioning rivers such as North Saint Vrain Creek provide a critical resource for understanding the basic processes operating along rivers and for evaluating a variety of land use effects on rivers.

Doug finished his work on the North Saint Vrain pool just in time. I returned to the creek the next summer as part of another project and discovered that a newly built beaver dam had flooded the entire reach and submerged the pool. The industrious beaver had chewed down all the dense shrubbery of willows that formerly had lined the stream banks. The creek looked so different that initially I wasn't sure I had the right spot. Beaver seldom occur along rivers heavily altered by human activities.

Beaver also have a tough time surviving in some parts of Rocky Mountain National Park. With the elimination of both predators and human hunting within the park, elk numbers climbed during the last three decades of the twentieth century. Three native species of willow *(Salix geyeriana, S. monticola, S. planifolia)* growing in wet meadows and stream areas provide an important food source for the elk, particularly during winter. The willows also provide food and habitat for many other species. An estimated 82 percent of bird breeding habitat in Colorado is in the riverside zone created by willow and other plants, and riverside areas have greater bird diversity than all other habitats combined. Riverside habitats provide key stopover zones for migratory birds, and bird species diversity and abundance increase with habitat complexity.

Beaver and willow are mutualists that promote each other's existence. By building dams that create ponds, beaver raise the local water table and create low-velocity zones where sediment settles and creates seedling beds for willows. Beaver benefit by eating high-protein willow leaves in summer and willow bark in winter.

The increasing number of elk is disrupting the relationship between beaver and willow. Willow cover has decreased by 20 percent or more during the past fifty to sixty years in areas of primary elk winter range. Heavily browsed willows assume a stunted, shrubby form that beaver do not eat. Predictably, beaver populations decline in areas used heavily by elk.

Population estimates for beaver in the Moraine Park section of Rocky Mountain National Park fell from 315 in 1939 to 6 in 1999. (Beaver trapping from 1941 to 1949 exacerbated the decline.) As beaver and their stabilizing dams disappear, and large elk herds destroy once-dense willow thickets, meandering streams in the park's meadows unravel, cutting straight, steep-sided trenches that lower the meadow's water table and increase fine sediment entering downstream reaches. Park scientists are now evaluating ways to reduce elk numbers, and they are establishing grazing exclosures that will allow beaver and willows to return.

.   .   .

Barring diversion into a tunnel or a ditch, the snowmelt of the high country continues to flow downstream along creeks that enlarge to become rivers. The upper portions of these creeks and rivers are mostly steep and rocky. The boulders may be rounded if they were left behind as a glacier melted, or angular if they fell from a bedrock cliff towering over the creek. Mainly they are composed of tough crystalline rock types that formed when the Rockies were being uplifted and intruded by hot magmas from deep within the Earth. If the valley is sufficiently steep, the rocks form vertical steps and the whole creek resembles a watery staircase. Where the slope of the valley grows gentler, the rocks form riffles that reflect the light in a thousand sparkles on sunny days.

Fish appear in the creeks as the channels grow larger. Endangered greenback cutthroat trout are present where a small waterfall prevents upstream migration of introduced trout species. Ouzels hunt insects along even the largest mountain rivers. The greatest number of aquatic insect species occurs in the transition zone, between nine thousand and four thousand feet elevation. Beaver build low dams of wood and sediment that pond water upstream and provide freshly deposited sediments where riverside trees and shrubs can take root along the edges of the pond. Bear, deer, and mountain lions come down to the rivers to drink. Willow and alder form a line of paler green between the river and the dark green conifer forest of the hillslopes.

At about eight thousand feet in elevation, the rivers cross the downstream-most deposits of the glaciers. These so-called terminal moraines are like small hills laid across the river valley that produce a flattening of the valley bottom immediately upstream, and a steeper, rocky zone immediately downstream. Once beyond this barrier, the rivers often have slightly gentler downstream slopes but can still steepen locally where a rockfall or a resistant bedrock ledge enters the channel. On the hillsides, spruce and fir give way to more pines, and the forest grows drier and more open.

Towns and roads are more likely to be present along the rivers below eight thousand feet in elevation, and contaminants associated with people enter the rivers in greater quantities. Sand and gravel spread on the roads to combat winter ice are carried by snowmelt into the adjacent rivers, where the sediment settles preferentially in pools. Metals such as zinc, cadmium, and lead drop onto the roads from vehicles and are carried into the rivers with the sediment. Volatile organic compounds, including benzene and gasoline additives, may enter the rivers from runoff, which can also carry household insecticides and other synthetic organochlorine compounds. Septic tanks and sewage treatment effluent introduce nutrients into the river, and the cobbles on the bed of the tea-brown streams become slick with mats of vividly green algae. These contaminants are all present in small amounts unless the river drains a mining region, and the water generally remains clean until it reaches the major urban and agricultural areas at the base of the mountains.

As the river descends below seventy-six hundred feet in elevation, the annual snowmelt flood can be supplemented by late-summer flash floods. Snowmelt floods are well behaved. The peak flow can be predicted from the size of the snowpack, and the peak is broad, lasting several days to weeks. In contrast, rainfall flash floods can be disastrous if humans or human structures are present along the river. Flash floods last at most a few hours, occur unpredictably, and bring huge volumes of water down a channel in a short time, creating a peak flow much larger than that of a snowmelt flood.

Summer thunderstorms occur all the way up to the Continental Divide. I have been caught in nasty storms near the timberline, when it seemed the world was about to end in black skies and crashing lightning. But the storm always passed in a matter of minutes, and within half an hour the sky was again blue and serene. Although air flowing from the Pacific can bring moisture, most of the summer moisture that reaches eastern Colorado comes with warm air flowing across the Great Plains from the Gulf of Mexico. As the air flows west, any topographic rise that forces it upward will cause the moisture to cool and condense, perhaps triggering a thunderstorm. The Rocky Mountains constitute the ultimate topographic rise, and air forced up the face of the Rockies can create violent thunderstorms. By the time they get to high elevations, fast-moving air masses retain relatively little moisture. But at the middle elevations of forty-five hundred to seventy-six hundred feet in elevation, the air masses can produce downpours that send torrents of water along the rivers. Thunderstorms usually move so fast that only the smallest creeks get enough rain to produce a flash flood. But occasionally the storms stall for minutes or hours, and then even the larger rivers can flood.

In July 1976 the middle elevations of the Big Thompson River received twelve inches of rainfall in two days. Average annual precipitation at those sites is about sixteen inches. In July 1997 parts of the Spring Creek drainage in Fort Collins received ten inches of rain in five and a half hours; average annual precipitation is fourteen inches. In each case, a massive flood resulted. The Big Thompson flood was estimated at 31,200 cubic feet per second at the canyon mouth, four times higher than the previous high discharge of 7,600 cubic feet per second in 1919. Parts of the highway along the river were ripped out, houses and power plants were destroyed, and 141 people were killed. The 1997 Spring Creek flood smashed homes, railways, and businesses and killed five people. Extreme flash floods like these completely reshape the river channel and the valley bottom, setting the template for the river's form until the next massive flood again rearranges things. From studies of sediments left by earlier floods and scars left on tree trunks along the channel by flood-borne debris, we estimate that such rearrangements occur every few hundred years along each river.

As with many processes in geomorphology, a lot happens at once during a flash flood, and then there is a long quiet period of little change. I missed the 1997 Spring Creek flood that thoroughly reconfigured a channel four miles from my house and flowed through the basement of my building at the university. I was working with a graduate student in the Yellowstone backcountry and had no idea a flood had occurred. Having spent the previous decade crawling into small caves above the Poudre River and other equally obscure spots in search of prehistoric flood sediments, I was disappointed that the city and the university cleared much of the flood debris before I returned to town. "You snooze, you lose" aptly describes geomorphic processes.

.   .   .

As mountain streams enter the elevations where flash flooding occurs, their appearance changes subtly. My graduate students Mike Grimm and Jon Pruess studied mountain streams from the headwaters to the base of the mountains, carefully mapping the elevations at which indicators of flash floods appear. At the lower elevations, piles of boulders where tributaries enter the main channel record flash floods and debris flows along the tributaries. Other signs of past flash floods—bouldery levees overgrown with trees beside the channel, or large scars high on the tree trunks— may be present along the river channel. Normal low stream flows and snowmelt floods do not have enough energy to reshape the bouldery flash-flood deposits, so these remain until the next flash flood or beyond.

Tall forests of ponderosa pines give way to scrubbier juniper and pinyon pine woodlands as the streams descend toward the mountain front. Oaks grow on the hillslopes, and river birch, vine maple, and cottonwood grow along the river channels. The boulders of the mountain channels give way to cobble and gravels as the streams enter a transition zone, where they flow between hills formed by sedimentary rocks tipped at steep angles as the Rockies were uplifted. Sandstones, shales, and limestones in these hills weather to boulders and cobbles that in turn quickly break down into sands, silts, and clays. The rivers become gentler, with fewer steep drops and more shallow riffles and runs. Deep pools form where the rivers bend at banks shored up with webs of cottonwood roots. The water temperature grows warmer.

· · ·

As a river enters the foothills and the plains, it may encounter a last remnant of mountainous topography where a finger of granite protrudes into the Great Plains and creates a small plateau and a short segment of deep canyon. The North Fork of the Cache la Poudre River crosses through a varied terrain of broad valleys with shortgrass prairie and deep granitic canyons, just before it enters the main Poudre River and flows east onto the Great Plains. In one of these canyons, the Nature Conservancy established the Phantom Canyon Preserve in 1989. The organization purchased the land from ranchers who ran cattle on the drier uplands and in the valley bottoms. Immediately upstream from the preserve, the North Poudre Irrigation Company operates Halligan Dam, built in 1910 to provide water storage for farmers and ranchers.

North Fork of the Cache la Poudre River in Phantom Canyon

Halligan Dam has an outlet valve that can release up to 140 cubic feet of water per second downstream into the North Fork, storing the rest of the river's flow above the dam. This is the maximum amount that can be taken into the company's downstream irrigation canal, which takes off from the river a few miles downstream. The company essentially uses the North Fork Poudre River as a path to get water from its reservoir to its canal. When the reservoir gets too high, water spills over the top of the dam. This doesn't hurt the dam, but, in the eyes of the irrigation company, it does waste water. The company prefers to have water flowing into its canal rather than continuing on down the river to the main Poudre, where other water users can get it.

The irrigation company has been operating Halligan Dam pretty much as it likes for several decades. This usually means storing as much water in the reservoir as possible, releasing not too much more than 140 cubic feet per second during the spring and summer, and then, toward the end of the growing season, drawing down the water stored in the reservoir. At the end of the growing season, flow is simply shut off below the dam and the reservoir is allowed to refill until the next spring. Shutting off the flow below the dam doesn't completely dry the channel of the North Fork Poudre River. Pools up to fourteen feet deep form along the river where a knob of bedrock protrudes from the canyon walls, constricting the river and causing scour of the streambed. The period of no flow during the winter simply means that these pools become isolated little ponds with no flowing surface water connecting them. When the Nature Conservancy established its preserve downstream from the dam, it purchased a small water right that allowed a trickle of water to flow between the pools throughout the dry autumn and winter.

The North Poudre Irrigation Company was also used to periodically releasing sediment from the reservoir during the autumn drawdown. The reservoir is gradually filling with sediment that would have continued downstream but is now trapped by the dam. This stored sediment means that the reservoir can store less water. Because the outlet-valve design on Halligan Dam permits sediment to be released only when the water level in the reservoir is quite low, if the irrigation company wants to flush sediment through the dam to increase reservoir storage capacity, it can do so only right before it shuts the flow down to almost nothing. This means that the sediment may be flushed through the dam, but it won't go very far downstream.

In late September 1996 the irrigation company flushed Halligan Reservoir of some of the sediment that had accumulated against the dam, and then shut off the flow. The sediment traveled about eight miles downstream, filling pools as it went. Pools immediately below the dam were completely filled. At three miles below the dam, pools were about half filled with a veneer of sand and silt that covered the original streambed. Continuing downstream, the sediment load petered out, until it was just a dusting of clay at eight miles downstream. In total, more than seven thousand cubic yards of sediment from the reservoir collected along the North Fork Poudre River.

The Nature Conservancy was not happy. Neither were the Colorado Division of Wildlife, Trout Unlimited, and many citizens of the region, once they heard about the sediment release. The Division counted four thousand dead fish. These were just the ones lying on top of or half buried in the mud of the now-dry channel. More fish were probably entombed beneath the mud. All of the fish were nonnative species of trout, but they had formed a self-sustaining population that earned a substantial revenue for the Conservancy when it auctioned fishing permits along the remote, uncrowded river canyon.

I first visited the canyon in early October, a few days after the sediment release, at the invitation of the Division of Wildlife. Phantom Canyon appears unexpectedly, as though a giant knife had been drawn across the nearly flat uplands, leaving an abrupt gash that cannot be seen until you are almost at the rim. The uplands undulate in broad, low hills covered with grasses and widely scattered pinyon and juniper trees. The Medicine Bow Mountains rise on the horizon to the west, white in winter above the bleached gold grasslands, and darker green in summer above the heat shimmers stirred by the grasslands' continual winds. Whenever I drive into the canyon from the nearest paved road, I almost always see a few pronghorn antelope or coyote, as well as different types of raptors working the winds.

Phantom Canyon is strikingly beautiful, but it stank on that hot October day. Together with fish biologists from the Colorado Division of Wildlife, I followed a trail that switchbacked steeply down from the rim to the green line formed by river birch and alder flanking the narrow river. Rotting fish were everywhere, their odor combining with the stink from all the decaying organic muck released from the reservoir. The biologists asked me to advise them on restoring the river and flushing the sediment downstream. Everyone wondered whether the sediment could be flushed in the next spring snowmelt, or whether it might take years to move downstream. And once the sediment was removed, how long might it take the aquatic insects and the fish to recover? Phantom Canyon hosts a thriving community of ouzels. What would the ouzels do if the sediment didn't move and the insects didn't come back? The sediment dump into the North Fork Poudre seemed to have destroyed the whole thriving aquatic ecosystem in a matter of hours.

I didn't know how long it might take the sediment to move downstream, but I initiated a study to observe what happened. With a continually changing ensemble of graduate students, I quickly chose segments of the stream to study and surveyed the channel condition and sediment infill. Crawling on our knees across the sticky sediment, we probed with long pieces of rebar until we hit the cobbles and boulders that formed the original pool bottom. Tracks of black bears and mountain lions were abundant in the gradually drying mud along the pools, and a variety of raptors gathered to feast, swiftly stripping fish carcasses to skeletons.

We dug into the sand and gravel forming the riffles to see how deeply the released silt and clay had penetrated, and how deeply the fine reservoir sediment would have to be flushed before fish could spawn again on the riffles. We celebrated after finding live fish in a pool three miles downstream.

Although disturbed by the unnecessary destruction of the North Fork's aquatic community, I was delighted to work in Phantom Canyon. I had not even known of its existence, despite its being only thirty-five miles north of my house. Now I found myself in a spectacular river environment, with the Nature Conservancy encouraging me to do research, and plenty of students eager to come out with me in order to see this canyon where visitors were normally allowed only limited access.

The winter of 1996–1997 was a good one for snow. The following spring saw high snowmelt flows of more than five hundred cubic feet per second, and these flushed most of the sediment from the North Fork pools and spread it downstream for many miles, without depositing enough at any one point to destroy insect or fish habitat. I was at the river two or three times a week while the snowmelt lasted, measuring flow velocity and sediment movement and periodically resurveying our study sites to see how the sediment storage was changing. Aquatic entomologists from the university measured how the pioneer species of insects recolonized the river during the next two years. The influx of fine organic detritus that came with the sediment released in 1996 caused a bloom of bright-green algal mats among the river-bottom cobbles normally swept clean by sand and gravel. After two to three years, the fish came back. The ouzels returned to nest along the river. The mix of aquatic species that had been present in the river ecosystem before the sediment release was gradually reestablished.

We did not grow complacent, however. The recovery of the river immediately following 1996 was heartening and impressive, but the timing was very lucky. The North Fork Poudre has not had a good year of high flow since 1997. In fact, the past few years have been so dry that in 2002 the river's flow never exceeded a hundred cubic feet per second. Colorado climate and stream flow are not so dependable that the irrigation company can count on good flows to correct for its bad judgment in releasing sediment.

Much of our research focused on pool dynamics because we realized that pools are the key controls on how a river such as the North Fork recovers following a massive sediment influx. Each pool acts as a small sediment reservoir, holding the sediment until all the upstream pools have been emptied. Sediment moves downstream as pools empty, like dominoes falling, and the amount of flow necessary to flush any particular pool depends on how many pools lie upstream. Our study of sediment movement and pool dynamics has turned out to be well timed, as interest in removing dams across the United States grows. The North Fork Poudre sediment flux provides a small case study of how rivers might respond when dams are removed and the sediment stored behind them for decades is once more released downstream.

Dams effectively plug rivers. They stop the passage of floods, the downstream movement of sediment, and the movement of fish both up- and downstream. They also stop the downstream drift of plant seeds. Dave Merritt used the North Fork to compare vegetation communities along mountain streams above and below dams. He modified the sediment collectors that my other students and I used, so that he could sample the drift of plant parts in the rivers. The samples collected downstream from dams showed a "seed shadow" that helps to create an impoverished streamside forest below the dam. Trees present below the dam also tend to be older. The dam dampens the occasional floods that historically ripped out mature trees, reconfiguring the channel and providing germination sites for new seedlings. Without these periodic disruptions, the streamside forest below the dam gradually ages and becomes more uniform, providing less wildlife habitat and species diversity.

. . .

As the rivers descending from the Rocky Mountains flow beyond the last granite outliers such as Phantom Canyon, they enter the Great Plains. Topographic relief decreases dramatically as the rivers flow across broad, gently undulating prairies. In places the rivers may intersect the underlying sedimentary rocks, but for the most part the channels are formed in fine gravels and sand deposited across the prairies for millions of years by wind and rivers. Rivers that originate in the mountains generally flow throughout the year, unless water diversions drain them completely. Tributaries that head on the plains are more likely to be ephemeral, flowing only during snowmelt or after rain.

The plains rivers experience the same types of episodic change as the mountain rivers. Large driftwood piles from a 1965 flood remain undisturbed along Plum Creek. Yet, during several years of field trips to an area just to the north, I have observed readily apparent changes in ephemeral streams cutting deep gullies. Understanding the different scales of change along rivers is part of the challenge of geomorphology.

River channels of the plains become broad and shallow downstream, shifting unpredictably back and forth across the broad valley bottoms as occasional floods recontour the channels. The

A small ephemeral channel in the Pawnee National Grassland of northeastern Colorado

water grows warmer, supporting species of fish different from the trout found in the mountains. Different species of insects live among the shifting sands. Riverside trees are relatively sparse and confined to small groves, rather than spread continuously along the river channels. The first people of European descent to reach rivers such as the Platte described them as "too thick to drink but too thin to plow" and "a mile wide and an inch deep." The early travelers took grainy black-and-white photographs in which the far bank of the river is lost on the horizon.

That was what the plains rivers looked like in the 1860s and 1870s. By 1900 they had begun to change dramatically, and today the larger rivers are hardly recognizable when compared to those photographs. The first settlers on the western plains needed more water to grow the crops they brought with them. These crops from the eastern and midwestern United States could not survive the dry years that periodic droughts brought to Colorado. Supplemental water could be pumped from underground, but this was expensive. It was cheaper to dig irrigation ditches and spread the spring snowmelt flows across the farm fields. Or better yet, they could build reservoirs to store those high flows for use throughout the growing season. As agricultural towns such as Fort Collins and Greeley attracted more and more people, an extensive network of dams and irrigation ditches thoroughly altered flow in the rivers.

The dams removed the annual flood peak that had shifted the sandy channels laterally and eroded the banks and kept them free of vegetation. The irrigation waters that were spread across farm fields raised the local water table and, together with the lack of flood flows, made it easier for trees to germinate and grow to maturity on the river banks. As the vegetation grew, roots held the stream banks together. Stems and leaves increased the hydraulic roughness during floods large enough to partially submerge the trees, and this slowed the current and allowed sediment to settle from suspension and accumulate along the river banks. The rivers grew narrower and deeper and began to meander. They changed from channels that had been fifteen hundred feet wide in the 1860s to channels three hundred feet wide by the early 1900s. What had been broad, unvegetated valley bottom and active channel became instead dense riverside forest.

These changes remain readily visible to anyone on an airplane flight over the Great Plains. Groves of cottonwoods that define the boundaries of a river's historical floodplain form a broad band alongside the narrow contemporary river. I am always the one with my window shade up during cross-country flights, because I still learn about landscapes from thirty-thousand feet up.

As the river channels underwent this metamorphosis, habitats changed. Water temperatures decreased in the deep, shaded channels. The annual cycle of high and low flow became more uniform. The constantly shifting sands stabilized. These changes stressed the native species that had adapted to the historical channel conditions, leading to the extinction of native fish. Species of fish, birds, and mammals more characteristic of wetter regions to the east invaded the western plains by following the now-forested river corridors. Among the native species most affected by changes in the plains rivers were migratory birds such as the sandhill crane and interior least tern. These birds use the broad floodplain and marsh habitats along the plains rivers for feeding and resting during their annual migration. As the river bottoms closed up with trees, migrating flocks were forced into ever smaller areas, exposing the birds to disease, predation, and starvation.

Of the thirty-eight native plains fish species known to exist in Colorado during the late 1800s, six are extinct as of 2008, and another thirteen are listed as endangered, threatened, or of special concern. Half the native fish species are either going or gone. Forty-three percent of Colorado's amphibian species are imperiled, and only 30 percent of the historical mussel populations remain today. Colorado and Ohio share the dubious distinction of being the only states to have an extinct species of mayfly, another indicator of altered rivers.

The metamorphosis of the big rivers and the introduction of exotic species are partly responsible for the extinctions. The drying of little rivers is also to blame. Pumping of groundwater for irrigated agriculture has lowered regional water tables to the point where pools that used to provide dry-season summer refuges along sandy creeks no longer exist.

Fish species that evolved to live in the Great Plains rivers are tough, mobile, and resilient. Unlike many fish, the native plains species tolerate huge annual fluctuations in the water temperature of their shallow streams, including summer temperatures as high as ninety-five degrees Fahrenheit. They also tolerate big swings in dissolved oxygen levels, and low dissolved oxygen. The fish adapt to the shifting sands of plains streambeds by laying buoyant eggs that are not crushed or buried by the sands. Fish such as the brassy minnow *(Hybognathus hankinsoni)* can outlast a drought in a pool with only a few square feet of water. But the fish need *some* water. They need pools in the summer dry season. They need flowing water during spring to connect pools so that adult fish can move upstream and reestablish populations that breed and release buoyant eggs to be dispersed downstream.

Historical changes along the plains rivers that have so stressed wildlife communities must be reversed to some extent to ease the pressure on threatened and endangered species. Conferences on water use and channel restoration are held regularly, but actions do not always follow talk. An extensive regional economy and many communities now depend on the water of the South Platte and other rivers to maintain modern lifestyles across the western plains. Land along the river bottoms must be purchased or designated for conservation easements. Water rights must be purchased and devoted to instream flows. And the deadly contaminants that now lace the soil, ground, and surface waters of agricultural and urban areas in the Great Plains must somehow be removed or contained before the rivers can be restored.

Water flowing from forested mountain drainage basins is

mostly clean and minimally affected by humans. But once the rivers flow through the first line of cities at the base of the mountains, water quality declines dramatically. Researchers in the U.S. Geological Survey's National Water Quality Assessment (NAWQA) program systematically sampled water quality along a transect from the Continental Divide to the eastern border with Nebraska during the years 1991 to 1995. NAWQA teams assessed water chemistry of surface and groundwaters, suspended sediment in streams, stream habitat degradation, and the integrity of aquatic communities in rivers. My graduate student Jill Minter and I worked with a NAWQA team, examining how the species composition of insect communities along plains streams reflected both the availability of different habitats and chemical contaminants. This was a rare time when I entered a river reluctantly. Looking upstream to the cow manure and urine churned into the mud at the edge of the stream bank, I resolved to shower very thoroughly when I got home. More insidious were the invisible contaminants in the river.

The NAWQA results reveal that, as rivers flow through urban communities, they become contaminated with everything from household chemicals and yard pesticides, to volatile organic carbons from gasoline and cleaning solvents. Wastewater treatment plants greatly increase the concentration of nitrates and phosphates, causing algal blooms and low levels of oxygen dissolved in the water. Presently banned compounds, such as the insecticide chlordane and polychlorinated biphenyls (PCBs)—the latter were widely used until 1979 in everything from carbon paper to ironing board covers—still appear in stream samples of sediment and fish tissue. River habitat and biological integrity still decline.

A particularly disturbing discovery made the front page of

Jill Minter working in the South Platte River near Greeley, Colorado

the *Denver Post* on October 3, 2004, in a story written by Theo Stein and Miles Moffeit. When biologists pulled white suckers *(Catostomus commersoni)*, fish native to Colorado, from the South Platte River at a point downstream from the Denver area's largest sewage plants, they found that many of the fish had both male and female reproductive parts, and that females far outnumbered males. The retired fisheries biologist John Woodling was quoted as stating, "This is the first thing I've seen as a scientist that really scared me." This type of endocrine disruption is not new, but this was the first time it had been detected in Colorado. By 2008, similar reports appeared regularly in Denver newspapers as biologists found intersex fish and fish loaded with pharmaceuticals and endocrine-disrupters in rivers and lakes throughout Colorado and the United States.

As I discuss in more detail in *Disconnected Rivers,* endocrine disrupters are a group of synthetic chemicals that mimic hormones and disrupt the development of many organisms, including humans. Theo Colborn and other scientists have warned for more than a decade of the increasing evidence that these chemicals are destroying the reproductive ability of organisms from invertebrates to mammals. Fish reproduce more quickly than humans, yet share some of the same basic biochemistry. I wonder about the implications for all the plains communities that pump their drinking water from shallow aquifers connected to this river.

Beyond the cities, the mountain-born rivers flow through only small towns until they reach the eastern border of Colorado. But between the urban corridor and the border lies an extensive agricultural region from which nitrates, salts, sediment, and pesticides leach into the rivers and groundwater. NAWQA scientists detected twenty-five different pesticides in the surface waters of the agricultural regions of eastern Colorado, and fifteen pesticides in groundwater samples. DDT, banned since 1972, was present in fish tissue and streambed sediment.

Irrigated agriculture occupies only 8 percent of the South Platte River basin but accounts for 71 percent of the water use. Onto these lands are dumped an estimated forty thousand tons of phosphorus and two hundred thousand tons of nitrogen each year. Water from half the groundwater wells sampled by NAWQA in agricultural areas had nitrate concentrations exceeding drinking-water standards established by the Environmental Protection Agency.

When I visit Greeley, an agricultural community downstream from my home at the base of the mountains, I see small kiosks selling purified water for twenty-five cents a gallon. The community of Fort Morgan, farther out on the plains, closed its contaminated wells and, at great expense, constructed a pipeline to bring drinking water directly from the mountainous portion of the Big Thompson River. Having fouled the limited water present on the dry plains, much of which originally came from the mountains, the citizens of the agricultural communities must now import clean water from the mountains by artificial means.

Thus ends the downstream journey of a snowflake falling in the Colorado Rockies. Having traveled more than two hundred miles as the crow flies, and much farther as the river flows, the snowflake has helped shape river courses from headwater cascades to broad, sandy channels on the plains. It has helped provide sustenance for insects, fish, birds, plants, and mammals at elevations from nearly fourteen thousand feet to four thousand feet. It has been continually recycled between these organisms and the river. Along the way, the melted snowflake has likely collected contaminants from human activities and passed these on to other organisms.

I used to think it was an open question as to whether we would so foul our water supplies that widespread deterioration in human health would result. Now I am convinced we have created this dire situation. The idea of poisoning the well—of poisoning the water source for a family or a community—has represented an evil and desperate deed throughout human history. American society is poisoning its own wells, literally and figuratively, but we have barely acknowledged this as a society or begun to effectively change the habits that cause the poisoning. For me, the poisoning of wells on the plains of Colorado is a warning sign that my society has not yet understood this landscape or how to live sustainably within it. We remain miners, here to extract what we can and then move on to new finds. But we are running out of new finds. Now we must remain here and deal with the poisons we have spread and the landscapes we have altered.

We do not lack the scientific or technical knowledge necessary to sustain water supplies, water quality, and stream ecosystems in Colorado. What we lack is the collective will to implement this knowledge by imposing penalties and offering incentives that alter existing behavior and patterns of resource use. We can change, if we choose.

Epilogue

Adventures in the Still *Unknown Interior of North America*

The truth I am trying to tell is a kind of waterhole never dried in any
drought. . . . I am helping to clear a track to unknown water.
Judith Wright

Written accounts of the western United States begin with Cabeza
de Vaca's 1542 *Adventures in the Unknown Interior of North
America*. Álvar Núñez Cabeza de Vaca was born in Andalusia
about 1490. He rose to distinction and, by royal appointment, was
made second-in-command to Pánfilo de Narváez in anticipation
of their expedition to Florida in 1527. Narváez abandoned Cabeza
de Vaca, along with three hundred of the expedition's cavalry and
infantry, on the Florida coast. Of all those men, only Cabeza de
Vaca and three companions found their way to Mexico City. For
eight years they traveled on foot through what is now Florida,
Alabama, Mississippi, Texas, New Mexico, Arizona, and down
the length of Mexico. They covered almost six thousand miles of
terra incognita. They were enslaved by some indigenous tribes,

virtually worshipped by others. They walked into new climates, landscapes, and fauna and flora. Through it all, Cabeza de Vaca observed and remembered.

Cabeza de Vaca was a tough man by the end of his journey, having walked all day and eaten only once each day, a spare evening meal. His habits matched those of the indigenous peoples he encountered. He wrote that Native Americans west of the Chiricahua Range in Arizona lived for a third of the year on nothing but dried desert herbs. This was all that the natives had been able to share with the Spaniards for seventeen *jornadas,* or day's journeys, before the Spaniards reached the corn-growing lands of the Opata tribes in Mexico.

Cabeza de Vaca's account provided tinder for the hopeful fires of Spanish treasure seekers. What he left unsaid, or hinted at, sent the expeditions of Coronado and Friar Marcos de Niza searching the Southwest for the mythical Seven Cities of Cíbola. Behind them came the earnest Jesuits, Franciscans, and Dominicans seeking souls to convert. Their extensive network of missions across the Southwest began a period of landscape transformation that accelerated with the settlement of Mexicans and then Anglo-Americans.

Most of these people perceived the landscape, and the people already inhabiting that landscape, through the dense filter of their own experience and expectations. Cabeza de Vaca's account suggests that he came to respect the ability of the indigenous peoples to live within a challenging physical environment. He was an exception. Many of the Spaniards, Mexicans, and Anglos who subsequently came to the American West put a low value on the abilities of people already inhabiting the region, and on the landscape itself, except as it could be altered to resemble a more desired land. The interior remained unknown to them because they never tried to understand the landscape or to live within it. Instead they imposed on it.

Others have protested eloquently against the dangers of seeing the western United States not as it is but as we wish it to be. Wallace Stegner wrote in *Where the Bluebird Sings to the Lemonade Springs:* "It is probably time we looked around us instead of looking ahead. We have no business, any longer, in being impatient with history. We need to know our history in much greater depth. . . . Neither the country nor the society we built out of it can be healthy until we stop raiding and running, and learn to be quiet part of the time, and acquire the sense not of ownership, but of belonging." I want to belong. Belonging grows from love and understanding and from recognition of limitations and responsibilities. These can only come from looking at the country and thinking about how its past and its present can together shape its future. This is knowing.

My first impression of the West was that of a vast playground. I grew up with American stories of individual cowboys and fur trappers heroically pitted against a challenging wilderness. I sought out such adventures when I came West and imagined myself alone in a pristine landscape. The whole West was "the last, best place," where time stopped and something still existed from before the twentieth century.

When I began to study geology and geomorphology, the West became a vast outdoor laboratory where I observed the workings of hillslopes and stream channels apart from human interference. I looked for drainage basins without roads, irrigation diversions, or housing subdivisions, and studied streams without human impacts. The West took on an elegiac quality as the last remnant of a largely vanished world that I was privileged to share. But the further I delved into both written and geomorphic history, the less these stories fit my earlier perceptions of the West.

The West now seems a trickster with a history that changes every time I think about it. The camels, horses, mastodons, and giant bison roaming the North American steppe may have been hunted to extinction by the first people to reach the continent. Or they may have vanished as climate and plant communities changed. Certainly the first people affected the landscape dramatically as they introduced different fire regimes and modified some areas for agriculture. The Europeans emigrating to America brought with them a broad array of plants and animals and cultural practices. They further modified the landscape. How should I even write of central Arizona, given its rapid changes in the past few decades?

Early Anglos found flowing streams and lush riverside wetlands that supported beaver and malarial mosquitoes in Arizona. Uplands away from the rivers appeared harsh and threatening. *Desert* was the word for any uninhabited wasteland: "the dreariest and most desolate country . . . positively hostile in its attitude towards every living thing except snakes, centipedes, and spiders" (Martha Summerhayes, 1874). People set out to spread water across the desert with irrigation systems. They envisioned the region as a garden that could be made to bloom with human help. They cited the biblical text saying that the desert shall be made to blossom as the rose. The climate that had been perceived as threatening to complexion and sanity became a boon to those with tuberculosis, arthritis, and other incurable ills. "Seek the advantages of the balmy desert climate to recuperate from the inroads of acute maladies" (1929 quote from an advertising brochure). By the late twentieth century, the desert epitomized the good life. Air conditioning and swimming pools mitigated the summer heat. The sun remained bright, the colors vibrant, and the air scented with flowers throughout the winter. Retirement communities beckoned with palms bending in the spring breezes along the golf course, backyard orange trees, and no snow to shovel. The retirement community in which my mother lives represents the ultimate ordered, engineered landscape, with its picturesquely curving streets along which neatly ordered houses sport yards of painted gravel and nonnative plants. "Live in a sunny, resort-like, master-planned community [that has] . . . everything under the sun" (2003 quote from an advertising brochure).

This is not the desert. It is a controlled landscape that contemporary Americans have superimposed on the desert, just as the thirteenth-century Hohokam superimposed irrigated fields and the nineteenth-century Mexicans superimposed herds of cattle. Have any humans known the desert? Have any humans been able to live in the desert for generations without reducing the ability of the land to support other forms of life and, ultimately, to support human life? I do not think so. Does that mean there is no point in trying to know the desert or any other Western interior? Again, I do not think so.

Past cultures have shaped the desert to their own needs and expectations to the extent they were able, leaving when other people displaced them or when the desert could no longer sustain them. Contemporary American culture has followed the same path. Now it is fast approaching the point at which it will no longer be sustainable, as water supplies are depleted or fouled and soil fertility is lost. There are other paths, but they require courage, determination, and understanding—*on the part of every-*

one, not just a few leaders. Many changes are needed. Voluntarily restricted population growth that will reduce the number of people sucking in a disproportionate share of resources is needed. Voluntary reduction in consumption of resources is needed. Voluntary restrictions on the extent of land used for cities and agriculture, and greater amounts of land left free from intensive human manipulation, are needed. All these changes must be voluntary. To wait until the changes are forced will be too late, for us and our land.

When I envision knowing this interior, and living within that knowledge, I do not think of a lone survivalist able to enter the landscape with only basic tools such as a knife and matches and wrest a living from a harsh environment. Nor do I think of a city of people imposing an imported lifestyle on the landscape and sucking resources from a far-flung region like some urban black hole. I envision a community of people who have made deliberate decisions to restrict their own numbers and consumption to a level that allows indigenous communities of plants and animals to persist through millennia.

I'm not sure that such a human community has ever existed on Earth. Even nonindustrialized peoples substantially alter the environments they inhabit; the extent and intensity of such alteration increase dramatically with industrialization. The fact that such a community has not existed does not mean that it *cannot* exist. Historical approximations exist, cases where people entering an area have altered the environment and caused the extinction of some species, but then reached an equilibrium that promised to retain existing nonhuman communities. In his book *Collapse,* Jared Diamond described sustainable societies ranging from those on the Pacific islands of Tikopia and Tonga, to those in the

New Guinea highlands, to the ancient Incan Empire of the central Andes, to the traditional Pueblo peoples of the southwestern United States. We have not yet reached such an equilibrium in the western United States. Habitat loss, pollution, and species extinction continue at frightening rates. Only a few rivers, such as the Yampa in Colorado and Utah and the Yellowstone in Montana, remain only slightly altered by humans and can serve as references against which to compare other rivers. I think we can reach an equilibrium, however, and at a higher level of environmental preservation than we are now tending toward. We will need to make informed, deliberate decisions rather than allow ignorant, piecemeal actions to determine our future. Until that time, we who live in the West will remain lost, adventuring in ignorance across unknown interiors.

My studies of landscape changes through time, and of the role humans have played in these changes, have helped me to understand the implications of human actions across time and space. By going westward, I have freed myself of some of my misconceptions about landscape. The collective efforts of hundreds of ecologists, geologists, and other researchers in the natural sciences have gradually assembled a story of how landscapes and ecosystems function, and of how humans influence this functioning. The challenge now is to integrate the insights arising from this research into the everyday choices made by individuals and by society in order to improve our collective ability to live sustainably in the American West. We can learn to define humans as an integral part of landscape.

The encouraging part of coming to understand interactions between humans and landscapes in the western United States has been my growing awareness of the resilience present in rivers and other natural systems. Despite the fact that the Denver Water Board, for example, removes nearly all the flow from headwater streams such as East Saint Louis Creek, aquatic insect communities persist below the diversions if some groundwater flow is present. When elk are excluded from riverside willows in Rocky Mountain National Park, the beaver return and build low dams across the rivers that reduce flow velocity and stabilize the eroding banks, as well as provide habitat diversity for aquatic insects and fish. Ephemeral streams cut vertical-walled arroyos into the deserts and dry grasslands of Arizona, but eventually those arroyos again fill with sediment and become shallow swales that provide seasonal wetlands. Wildfires that kill the forest covering steep, granitic hillslopes trigger massive debris flows that fill stream channels and cause out-of-channel floods, but over a few years' time the vegetation regrows on the slopes and stream flows again scour out pools along the channel.

Not everything, however, can be restored. Extinction of a species is irreversible. Contamination of groundwater or streambed sediments by synthetic chemicals that require centuries to break down is essentially permanent in the time span of contemporary societies. There are limits to the resilience of streams, and sometimes we do not have sufficient knowledge to realize when these limits are exceeded. I think that a part of living sustainably within the West will be to increase our caution about not exceeding limits.

Nearly thirty years after I took up residence in the arid Intermountain West, the vivid colors and the long views that provide a sense of vast spaces uncluttered by buildings and infrastructure continue to define my sense of place. I now understand that the absence of infrastructure does not translate to the absence of human effects on the landscape, but I also understand that the landscape continually absorbs and responds to changes, and that most landscapes have some resilience in responding to the impoverishing and homogenizing effects of many land uses.

I once came across a curious trail in the snow. A mouse had strayed from its established routes and become lost. Beginning with a wide circle, the mouse moved in a concentrically narrowing spiral, its tracks clearly recorded in the deep powder. At the center of the spiral the mouse lay dead. Sometimes I feel that our society is moving in such a spiral, blindly following the path initially chosen. But I also believe that an understanding of our history, of our own interior of perceptions and desires, and of our land's interior, will preserve us from the mouse's fate.

BIBLIOGRAPHIC SOURCES
AND FURTHER READING

THE WESTERN RESERVE

Brown, Dee. *The Gentle Tamers: Women of the Old Wild West.* Lincoln: University of Nebraska Press, 1958.

Carson, Rachel. *Silent Spring.* New York: Houghton Mifflin, 1962.

Crèvecoeur, J. Hector St. John de. *Letters from an American Farmer and Sketches of Eighteenth-Century America.* 1782. Reprint, New York: Penguin Classics, 1981.

Darwin, Charles. *The Voyage of the Beagle.* 1860. Reprint, Garden City, NY: Doubleday, 1962.

De Voto, Bernard. *The Western Paradox: A Conservation Reader.* New Haven, CT: Yale University Press, 2000.

Goetzmann, William H. *Exploration and Empire: The Explorer and the Scientist in the Winning of the American West.* New York: Norton, 1966.

Goetzmann, William H., and William N. Goetzmann. *The West of the Imagination.* New York: Norton, 1986.

Kazin, Alfred. *A Writer's America: Landscape in Literature.* New York: Knopf, 1988.

Knepper, George W. *Ohio and Its People.* Kent, OH: Kent State University Press, 1989.

Luchetti, Cathy, and Carol Olwell. *Women of the West.* Saint George, UT: Antelope Island Press, 1982.

McNeill, William H. *The Great Frontier: Freedom and Hierarchy in Modern Times.* Princeton, NJ: Princeton University Press, 1983.

National Park Service. Division of Publications. *Exploring the American West, 1803–1879.* Handbook 116. Washington, DC: U.S. Department of Interior, 1982.

Newcombe, Jack, ed. *Travels in the Americas.* New York: Weidenfeld and Nicolson, 1989.

Stewart, Elinore Pruitt. *Letters of a Woman Homesteader.* 1913. Reprint, Lincoln: University of Nebraska Press, 1961.

Twain, Mark. *Roughing It.* London: G. Routledge, 1882.

Unruh, John David, Jr. *The Plains Across: The Overland Emigrants and the Trans-Mississippi West, 1840–60.* Chicago: University of Illinois Press, 1979.

A SENSE OF SPACE

Anaya, Rudolfo. *Bless Me, Ultima.* Berkeley, CA: Tonatiuh International, 1972.

Austin, Mary. *The Land of Little Rain.* 1903. Reprint, Albuquerque: University of New Mexico Press, 1974.

———. *The Land of Journey's Ending.* 1924. Reprint, Tucson: University of Arizona Press, 1983.

Cather, Willa S. *Death Comes for the Archbishop.* New York: Knopf, 1950.

Comeaux, Malcolm L. *Arizona: A Geography.* Boulder, CO: Westview Press, 1981.

Dobyns, Henry F. *From Fire to Flood: Historic Human Destruction of Sonoran Desert Riverine Oases.* Socorro, NM: Ballena Press, 1981.

Dohrenwend, John C. "Basin and Range." In *Geomorphic Systems of North America,* ed. W. L. Graf, pp. 303–342. Boulder, CO: Geological Society of America, 1987.

Hall, Sharlot M. *Poems of a Ranch Woman.* Phoenix, AZ: Sharlot Hall Historical Society, 1953.

Hastings, James R., and Ray M. Turner. *The Changing Mile: An Ecologi-*

cal Study of Vegetation Change with Time in the Lower Mile of an Arid and Semiarid Region.* Tucson: University of Arizona Press, 1965.

Horgan, Paul. *Great River: The Rio Grande in North American History.* Austin: Texas Monthly Press, 1984.

Krech, Shepard. *The Ecological Indian: Myth and History.* New York: Norton, 1999.

Krutch, Joseph W. *The Desert Year.* New York: Viking, 1952.

———. *The Voice of the Desert: A Naturalist's Interpretation.* New York: Morrow Quill Paperbacks, 1955.

Martin, Paul S. *The Last 10,000 Years: A Fossil Pollen Record of the American Southwest.* Tucson: University of Arizona Press, 1963.

McPhee, John. *Basin and Range.* New York: Farrar, Straus, Giroux, 1981.

Meinig, D. W. *Southwest: Three Peoples in Geographical Change, 1600–1970.* New York: Oxford University Press, 1971.

Nabhan, Gary Paul. *The Desert Smells Like Rain: A Naturalist in Papago Indian Country.* San Francisco: North Point Press, 1982.

Niethammer, Carolyn. *Daughters of the Earth: The Lives and Legends of American Indian Women.* New York: Collier, 1977.

Opler, Morris E., ed. *Grenville Goodwin among the Western Apache: Letters from the Field.* Tucson: University of Arizona Press, 1973.

Preston, Douglas. *Cities of Gold.* New York: Simon and Schuster, 1992.

Rea, Amadeo M. *Once a River: Bird Life and Habitat Changes on the Middle Gila.* Tucson: University of Arizona Press, 1983.

Spicer, Edwin H. *Cycles of Conquest: The Impact of Spain, Mexico, and the United States on the Indians of the Southwest, 1533–1960.* Tucson: University of Arizona Press, 1962.

Summerhayes, Martha. *Vanished Arizona: Recollections of the Army Life of a New England Woman.* 1911. Reprint, Lincoln: University of Nebraska Press, 1979.

Van Dyke, John C. *The Desert.* 1901. Reprint, Salt Lake City, UT: Peregrine Smith, 1980.

Waters, Frank. *People of the Valley.* Denver: Sage Books, 1941.

———. *The Colorado.* 1946. Reprint, Athens, OH: Ohio University Press, Swallow Press, 1984.

Young, Herbert V. *Water by the Inch: Adventures of a Pioneer Family on an Arizona Desert Homestead.* Flagstaff, AZ: Northland Press, 1983.

RIVER DAYS: PARADISE FOUND

Ambler, J. Richard. *The Anasazi: Prehistoric People of the Four Corners Region.* Flagstaff: Museum of Northern Arizona Press, 1977.

Beal, Merrill D. *Grand Canyon: The Story behind the Scenery.* Las Vegas, NV: KC Publications, 1967.

Dellenbaugh, Frederick S. *The Romance of the Colorado River.* New York: G. P. Putnam's Sons, 1903.

———. *A Canyon Voyage.* New Haven, CT: Yale University Press, 1908.

Everhart, Ronald E. *Glen Canyon–Lake Powell: The Story behind the Scenery.* Las Vegas, NV: KC Publications, 1983.

Fowler, Don D., Robert C. Euler, and Catherine S. Fowler. *John Wesley Powell and the Anthropology of the Canyon Country.* U.S. Geological Survey Professional Paper 670. Washington, DC: U.S. Government Printing Office, 1969.

Grattan, Virginia L. *Mary Colter: Builder upon the Red Earth.* Flagstaff, AZ: Northland Press, 1980.

Krutch, Joseph W. *Grand Canyon: Today and All Its Yesterdays.* New York: Morrow, 1958.

Leydet, Francois. *Time and the River Flowing: Grand Canyon.* Sierra Club: Sierra Club Books; New York: Ballantine, 1964.

Péwé, Troy L. *Colorado River Guidebook: Lees Ferry to Phantom Ranch.* Phoenix, AZ: Sims Printing Company, 1968.

Powell, John Wesley. *The Exploration of the Colorado River and Its Canyons.* 1895. Reprint, New York: Dover, 1961.

Stegner, Wallace. *Beyond the Hundredth Meridian: John Wesley Powell and the Second Opening of the West.* Lincoln: University of Nebraska Press, 1954.

Trimble, Stephen. *The Bright Edge: A Guide to the National Parks of the Colorado Plateau.* Flagstaff: Museum of Northern Arizona Press, 1979.

Zwinger, Ann. *Run, River, Run: A Naturalist's Journey down One of the Great Rivers of the American West.* Tucson: University of Arizona Press, 1975.

RIVER DAYS: PARADISE LOST

Abbey, Edward. *Down the River.* New York: E.P. Dutton, 1982.

Collier, Michael, Robert H. Webb, and John C. Schmidt. *Dams and Rivers: Primer on the Downstream Effects of Dams.* U.S. Geological Survey Circular 1126, 1996.

Fradkin, Philip L. *A River No More: The Colorado River and the West.* Tucson: University of Arizona Press, 1981.

Graf, William. "Fluvial Adjustments to the Spread of Tamarisk in the Colorado Plateau Region." *Geological Society of America Bulletin* 89 (1978): 1491–1501.

Kieffer, Susan W. "The 1983 Hydraulic Jump in Crystal Rapid: Implications for River-Running and Geomorphic Evolution in the Grand Canyon." *Journal of Geology* 93 (1983): 385–406.

Leopold, Aldo. *A Sand County Almanac, with Essays on Conservation from Round River.* New York: Ballantine, 1966.

Melis, Theodore S., Robert H. Webb, Peter G. Griffiths, and T.W. Wise. *Magnitude and Frequency Data for Historic Debris Flows in Grand Canyon National Park and Vicinity, Arizona.* U.S. Geological Survey Water-Resources Investigations Report 94–4214. Tucson, AZ: Department of the Interior, U.S. Geological Survey, 1995.

O'Connor, Jim E., Lisa L. Ely, Ellen E. Wohl, Larry E. Stevens, Theodore S. Melis, V.S. Kale, and V.R. Baker. "A 4500-Year-Record of Large Floods on the Colorado River in the Grand Canyon, Arizona." *Journal of Geology* 102 (1994): 1–9.

Powell, John Wesley. *The Exploration of the Colorado River and Its Canyons.* 1895. Reprint, New York: Dover, 1961.

Stephens, Hal G., and Eugene M. Shoemaker. *In the Footsteps of John Wesley Powell: An Album of Comparative Photographs of the Green and Colorado Rivers, 1871–72 and 1968.* Boulder, CO: Johnson, 1987.

Sykes, Godfrey. *The Colorado Delta.* Port Washington, NY: Kennikat Press, 1937.

Webb, Robert H. *Grand Canyon, a Century of Change: Rephotography of the 1889–1890 Stanton Expedition.* Tucson: University of Arizona Press, 1996.

Whitman, Walt. *Leaves of Grass.* 1855. Reprint, New York: Penguin Books, 2005.

THE DELICATE STRENGTH OF ROCK

Abbey, Edward. *Desert Solitaire: A Season in the Wilderness.* New York: Ballantine, 1968.

Bezy, John. *Bryce Canyon: The Story behind the Scenery.* Las Vegas, NV: KC Publications, 1980.

Bolton, Herbert H. *Pageant in the Wilderness: The Story of the Escalante Expedition to the Interior Basin, 1776.* Salt Lake City: Utah Historical Society, 1950.

Brooks, Juanita. *Quicksand and Cactus: A Memoir of the Southern Mormon Frontier.* Salt Lake City, UT: Howe Brothers, 1982.

Cather, Willa S. *The Professor's House.* New York: Grossett and Dunlap, 1925.

———. *The Song of the Lark.* London: J. Cape, 1932.

Gilbert, Bill. *Westering Man: The Life of Joseph Walker.* Norman: University of Oklahoma Press, 1983.

Graf, William L., Richard Hereford, Julie Laity, and Richard A. Young. "The Colorado Plateau." In *Geomorphic Systems of North America,* ed. William L. Graf, pp. 259–302. Boulder, CO: Geological Society of America, 1987.

Hall, Sharlot M. *Poems of a Ranch Woman.* Phoenix, AZ: Sharlot Hall Historical Society, 1953.

———. *Sharlot Hall on the Arizona Strip: A Diary of a Journey through Northern Arizona in 1911,* ed. G. C. Crampton. Flagstaff, AZ: Northland Press, 1975.

Lavendar, David. *One Man's West.* New York: Doubleday and Company, 1943.

Maxwell, Margaret F. *A Passion for Freedom: The Life of Sharlot Hall.* Tucson: University of Arizona Press, 1982.

McClure, Grace. *The Bassett Women.* Athens, OH: Ohio University Press, Swallow Press, 1985.

Meloy, Ellen. *Raven's Exile: A Season on the Green River.* New York: Henry Holt, 1994.

Morgan, Dale L. *Jedediah Smith and the Opening of the West.* Lincoln: University of Nebraska Press, 1953.

Osmundson, D. B., R. J. Ryel, V. L. Lamarra, and J. Pitlick. "Flow-Sediment-Biota Relations: Implications for River Regulation Effects on Native Fish Abundance." *Ecological Applications* 12 (2002): 1719–1739.

Rusho, W. L. *Everett Ruess: A Vagabond for Beauty.* Salt Lake City, UT: Peregrine Smith, 1983.

Stegner, Wallace. *Mormon Country.* Lincoln: University of Nebraska Press, 1942.

———. *The Gathering of Zion: The Story of the Mormon Trail.* 1964. Reprint, Salt Lake City, UT: Westwater, 1981.

Warner, Ted J., ed. *The Dominguez-Escalante Journal: Their Expedition through Colorado, Utah, Arizona, and New Mexico in 1776.* Trans. A. Chavez. 1776. Reprint, Provo, UT: Brigham Young University Press, 1976.

Webb, Robert H., and Sara L. Rathburn. "Paleoflood Hydrologic Research in the Southwestern United States." *Transportation Research Record* 1201 (1989): 9–21.

Wick, Edmund J. "Physical Processes and Habitat Critical to the Endangered Razorback Sucker on the Green River, Utah." PhD diss., Colorado State University, Fort Collins, 1997.

Wohl, Ellen E., Douglas M. Thompson, and Andrew J. Miller. "Canyons with Undulating Walls." *Geological Society of America Bulletin* 111 (1999): 949–959.

Zwinger, Ann. *Wind in the Rock: The Canyonlands of Southeastern Utah.* Tucson: University of Arizona Press, 1978.

THE WESTERN RAMPART

Armstrong, David M. *Distribution of Mammals in Colorado.* Monograph of the Museum of Natural History, No. 3. Lawrence: University of Kansas, 1972.

Baars, D. L., B. L. Bartleson, C. E. Chapin, B. F. Curtis, R. H. de Voto, J. R. Everett, R. C. Johnson, C. M. Molenaar, F. Peterson, C. J. Schenk, J. D. Love, I. S. Merin, P. R. Rose, R. T. Ryder, N. B. Waechter, and L. A. Woodward. "Basins of the Rocky Mountain Region." In *Sedimentary Cover—North American Craton; United States,* ed. Lawrence L. Sloss, pp. 109–220. Boulder, CO: Geological Society of America, 1988.

Benedict, James B. "Prehistoric Man and Climate: The View from Timberline." In *Quaternary Studies,* ed. R. P. Suggate and M. M. Creswell, pp. 67–74. Wellington: Royal Society of New Zealand, 1975.

———. *Old Man Mountain: A Vision Quest Site in the Colorado High Country.* Research Report No. 4. Ward, CO: Center for Mountain Archeology, 1985.

Benedict, James B., and B. L. Olson. *The Mount Albion Complex: A Study of Prehistoric Man and the Altithermal.* Research Report No. 1. Ward, CO: Center for Mountain Archeology, 1978.

Berman, A. E., D. Poleschook, and T. E. Dimelow. "Jurassic and Cretaceous Systems of Colorado." In *Colorado Geology,* ed. Harry C. Kent and Karen W. Porter, pp. 111–128. Denver: Rocky Mountain Association of Geologists, 1980.

Bird, Isabella L. *A Lady's Life in the Rocky Mountains.* 1878. Reprint, Norman: University of Oklahoma Press, 1960.

Buchholtz, Charles W. *Rocky Mountain National Park: A History.* Boulder: Colorado Associated University Press, 1983.

Debo, Angie. *A History of the Indians of the United States.* Norman: University of Oklahoma Press, 1970.

DeLoria, Vine, Jr., and C. M. Lytle. *American Indians, American Justice.* Austin: University of Texas Press, 1983.

Ellis, Anne. *The Life of an Ordinary Woman.* Boston: Houghton Mifflin, 1929.

———. *Plain Anne Ellis.* Boston: Houghton Mifflin, 1931.

———. *Sunshine Preferred.* Boston: Houghton Mifflin, 1934.

Frémont, John Charles. *Report of the Exploring Expedition to the Rocky Mountains in the Year 1842, and to Oregon and North California in the Years 1843–'44.* Washington, DC: Gales and Seaton, 1845.

Garland, Hamlin. *A Daughter of the Middle Border.* New York: Macmillan, 1929.

Grant, Marcus P. "A Fluted Projectile Point from Lower Poudre Canyon, Larimer County, Colorado." *Southwestern Lore* 54 (1988): 4–7.

Greeley, Horace. *An Overland Journey, from New York to San Francisco, in the Summer of 1859.* 1860. Reprint, Ann Arbor, MI: University Microfilms, 1966.

Madole, Richard F., William C. Bradley, D. S. Loewenherz, Dale F. Ritter, N. W. Rutter, and C. E. Thorn. "Rocky Mountains." In *Geo-*

morphic Systems of North America, ed. W. L. Graf, pp. 211–257. Boulder, CO: Geological Society of America, 1987.

Marsh, Charles S. *People of the Shining Mountains.* Boulder, CO: Pruett, 1982.

Mejia-Navarro, Mario. "Integrated Planning Decision Support System Incorporating Geological Hazards and Risk Assessment." PhD diss., Colorado State University, Fort Collins, 1995.

Mejia-Navarro, Mario, Ellen E. Wohl, and D. Oaks. "Geological Hazards, Vulnerability, and Risk Assessment Using GIS: Model for Glenwood Springs, Colorado." *Geomorphology* 10 (1994): 331–354.

Muir, John. *Our National Parks.* Boston: Houghton Mifflin, 1901.

Olson, Richard, and Wayne A. Hubert. *Beaver: Water Resources and Riparian Habitat Manager.* Laramie: University of Wyoming, 1994.

Pike, Zebulon M. *Exploratory Travels through the Western Territories of North America: Comprising a Voyage from St. Louis, on the Mississippi, to the Source of That River, and a Journey through the Interior of Louisiana, and the North-Eastern Provinces of New Spain.* 1811. Reprint, Denver: W. H. Lawrence, 1889.

Robertson, Janet. *The Magnificent Mountain Women: Adventures in the Colorado Rockies.* Lincoln: University of Nebraska Press, 1990.

Ruxton, George F. *Wild Life in the Rocky Mountains.* 1916. Reprint, New York: Macmillan, 1926.

———. *Life in the Far West,* ed. L. R. Hafen. 1849. Reprint, Norman: University of Oklahoma Press, 1951.

Sandoz, Mari. *Cheyenne Autumn.* New York: Avon, 1953.

———. *The Beaver Men: Spearheads of Empire.* Lincoln: University of Nebraska Press, 1964.

———. *Crazy Horse: The Strange Man of the Oglalas.* Lincoln: University of Nebraska Press, 1992.

Solnit, Rebecca. *Savage Dreams: A Journey into the Landscape Wars of the American West.* Berkeley: University of California Press, 1994.

———. *Wanderlust: A History of Walking.* New York: Penguin, 2000.

Stearn, Colin W., Robert L. Carroll, and Thomas H. Clark. *Geological Evolution of North America.* 3rd ed. New York: Wiley and Sons, 1979.

Stone, Irving. *Men to Match My Mountains.* New York: Berkley, 1956.

Taylor, Bayard. *Colorado: A Summer Trip.* 1867. Reprint, Niwot: University Press of Colorado, 1989.

Ubbelohde, Carl, Maxine Benson, and Duane A. Smith. *A Colorado History.* 6th ed. Boulder, CO: Pruett, 1988.

Veblen, Thomas T., and Diane C. Lorenz. *The Colorado Front Range: A Century of Ecological Change.* Salt Lake City: University of Utah Press, 1991.

Wohl, Ellen E. *Virtual Rivers: Lessons from the Mountain Rivers of the Colorado Front Range.* New Haven, CT: Yale University Press, 2001.

Zwinger, Ann. *Beyond the Aspen Grove.* Tucson: University of Arizona Press, 1981.

WHERE THE WINDS LIVE

Abbey, Edward. *The Fool's Progress: An Honest Novel.* New York: Henry Holt, 1988.

Bierhorst, John. *The Mythology of North America.* New York: William Morrow, 1985.

Boorstin, Daniel J. *The Americans: The National Experience.* New York: Vintage, 1965.

Borland, H. *High, Wide, and Lonesome.* New York: J. B. Lippincott, 1956.

Cassells, Edith S. *The Archaeology of Colorado.* Boulder, CO: Johnson, 1983.

Cather, Willa S. *O Pioneers!* New York: Houghton Mifflin, 1913.

———. *The Song of the Lark.* London: J. Cape, 1932.

Cronon, William. *Nature's Metropolis: Chicago and the Great West.* New York: Norton, 1991.

De Voto, Bernard. *The Year of Decision: 1846.* Boston: Houghton Mifflin, 1942.

———. *Across the Wide Missouri.* Boston: Houghton Mifflin, 1947.

———. *De Voto's West: History, Conservation, and the Public Good,* ed. Edward K. Mueller. Athens, OH: Ohio University Press, Swallow Press, 2005.

Donaldson, James C., and Logan MacMillan. "Oil and Gas: History of Development and Principal Fields in Colorado." In *Colorado Geology,* ed. Harry C. Kent and Karen W. Porter, pp. 175–189. Denver: Rocky Mountain Association of Geologists, 1980.

Ehrlich, Gretel. *The Solace of Open Spaces.* New York: Penguin, 1985.

Eighmy, Jeffrey L. *Colorado Plains Prehistoric Context for Management of Prehistoric Resources of the Colorado Plains.* Denver: State Historical Society of Colorado, 1984.

Erdoes, Richard, and Alfonso Ortiz, eds. *American Indian Myths and Legends.* New York: Pantheon, 1984.

Ewers, John C., Marsha V. Gallagher, David C. Hunt, and Joseph C. Porter. *Views of a Vanishing Frontier.* Lincoln: University of Nebraska Press, 1984.

Flores, Dan. *Caprock Canyonlands.* Austin: University of Texas Press, 1990.

Frazier, Ian. *Great Plains.* New York: Penguin, 1989.

Garland, Hamlin. *A Son of the Middle Border.* New York: Macmillan, 1924.

———. *A Daughter of the Middle Border.* New York: Macmillan, 1929.

Gregg, Josiah. *The Commerce of the Prairies.* 1844. Reprint, Lincoln: University of Nebraska Press, 1967.

Hungry Wolf, Beverly. *The Ways of My Grandmothers.* New York: Quill, 1980.

James, Edwin. *Account of an Expedition from Pittsburgh to the Rocky Mountains.* 2 vols. 1822–23. Reprint, Ann Arbor, MI: University Microfilms, 1966.

Lecompte, Janet. *Pueblo, Hardscrabble, Greenhorn: Society on the High Plains, 1832–1856.* Norman: University of Oklahoma Press, 1978.

McMurtry, Larry. *Buffalo Girls.* New York: Simon and Schuster, 1990.

Muhs, Daniel R. "Age and Paleoclimatic Significance of Holocene Sand Dunes in Northeastern Colorado." *Annals of the Association of American Geographers* 75 (1985): 566–582.

Osterkamp, Waite R., Mark M. Fenton, Thomas C. Gustavson, Richard F. Hadley, Vance T. Holliday, Roger B. Morrison, and Terrence J. Toy. "The Great Plains." In *Geomorphic Systems of North America,* ed. William L. Graf, pp. 163–210. Boulder, CO: Geological Society of America, 1987.

Parkman, Francis. *The Oregon Trail.* 1847. Reprint: Garden City, NY: Garden City Publishing, 1948.

Peattie, Donald C. *A Prairie Grove.* New York: Simon and Schuster, 1938.

Popper, Deborah, and Frank Popper. "The Great Plains: From Dust to Dust." *Planning* 53 (1997).

Powell, John Wesley. *Report on the Lands of the Arid Region of the United States,* ed. Wallace Stegner. 1878. Reprint, Cambridge, MA: Harvard University Press, Belknap Press, 1962.

Rolvaag, Ole E. *Giants in the Earth.* New York: Blue Ribbon Books, 1938.

Sandoz, Mari. *Old Jules.* Lincoln: University of Nebraska Press, 1935.

―――. *Love Song to the Plains.* Lincoln: University of Nebraska Press, 1961.

Scott, Michael L., Jonathan M. Friedman, and Gregor T. Auble. "Fluvial Process and the Establishment of Bottomland Trees." *Geomorphology* 14 (1996): 327–339.

Stegner, Wallace. *The Big Rock Candy Mountain.* New York: Penguin, 1938.

―――. *Beyond the Hundredth Meridian: John Wesley Powell and the Second Opening of the West.* Lincoln: University of Nebraska Press, 1954.

―――. *Wolf Willow: A History, a Story, and a Memory of the Last Plains Frontier.* Lincoln: University of Nebraska Press, 1962.

―――, ed. *The American Novel: From James Fenimore Cooper to William Faulkner.* New York: Basic Books, 1965.

Svingen, Orlan J. *The Northern Cheyenne Indian Reservation, 1877–1900.* Niwot: University Press of Colorado, 1993.

Sykes, Hope W. *Second Hoeing.* New York: G. P. Putnam's Sons, 1935.

Webb, Walter Prescott. *The Great Plains.* Boston: Ginn and Company, 1931.

―――. *The Great Frontier.* London: Secker and Warburg, 1953.

COLORADO BURNING

Hansen, Andrew J., et al. "Ecological Causes and Consequences of Demographic Change in the New West." *BioScience* 52 (2002): 151–162.

Maclean, John. *Fire and Ashes: On the Front Line of American Wildfire.* New York: Henry Holt, 2003.

Pyne, Stephen J. *Fire in America: A Cultural History of Wildland and Rural Fire.* Princeton, NJ: Princeton University Press, 1982.

Wohl, Ellen E., and Philip A. Pearthree. "Debris Flows as Geomorphic Agents in the Huachuca Mountains of Southeastern Arizona." *Geomorphology* 4 (1991): 273–292.

LET IT SNOW!

Clifford, Hal. *Downhill Slide: Why the Corporate Ski Industry Is Bad for Skiing, Ski Towns, and the Environment.* San Francisco: Sierra Club Books, 2002.

EQUIFINALITY

De Voto, Bernard. *De Voto's West: History, Conservation, and the Public Good,* ed. Edward K. Mueller. Athens, OH: Ohio University Press, Swallow Press, 2005.

Gilbert, Grove K. *Report on the Geology of the Henry Mountains.* Washington, DC: Government Printing Office, 1877.

———. *Lake Bonneville.* Monographs of the U.S. Geological Survey. Vol. 1. Washington, DC: Government Printing Office, 1890.

Graf, William L. "The Arroyo Problem—Palaeohydrology and Palaeohydraulics in the Short Term." In *Background to Palaeohydrology,* ed. K. J. Gregory, pp. 279–302. Chichester, U.K.: Wiley and Sons, 1983.

Hall, Sharlot A. "Channel Trenching and Climatic Change in the Southern United States Great Plains." *Geology* 18 (1990): 342–345.

Leopold, Luna B. "Reversal of Erosion Cycle and Climatic Change." *Quaternary Research* 6 (1976): 557–562.

McPhee, John. *Rising from the Plains.* New York: Farrar, Straus, Giroux, 1986.

Patton, Peter C., and Stanley A. Schumm. "Gully Erosion, Northwestern Colorado: A Threshold Phenomenon." *Geology* 3 (1975): 88–90.

Powell, John Wesley. *Report on the Lands of the Arid Region of the United States,* ed. Wallace Stegner. 1878. Reprint, Cambridge, MA: Harvard University Press, Belknap Press, 1962.

Pyne, Stephen J. *Grove Karl Gilbert: A Great Engine of Research.* Austin: University of Texas Press, 1980.

Schumm, Stanley A., and Richard F. Hadley. "Arroyos and the Semiarid Cycle of Erosion." *American Journal of Science* 255 (1957): 161–174.

Schumm, Stanley A., and R. F. Parker. "Implications of Complex Response of Drainage Systems for Quaternary Alluvial Stratigraphy." *Nature* 243 (1973): 99–100.

Webb, Robert H., Spence S. Smith, and V. Alexander McCord. "Historic Channel Change of Kanab Creek, Southern Utah and Northern Arizona." Monograph No. 9. Grand Canyon, AZ: Grand Canyon Natural History Association, 1991.

WHAT IS NATURAL?

Foster, David, G. Motzkin, J. O'Keefe, E. Boose, D. Orwig, J. Fuller, and B. Hall. "The Environmental and Human History of New England." In *Forests in Time: The Environmental Consequences of 1,000 Years of Change in New England,* ed. David R. Foster and John D. Aber, pp. 43–100. New Haven, CT: Yale University Press, 2004.

Hall, Marcus. *Earth Repair: A Transatlantic History of Environmental Restoration.* Charlottesville: University of Virginia Press, 2005.

Jaquette, Christopher D. "Historical Analysis and Sediment Budget Development of the North Fork of the Gunnison River, Colorado." Master's thesis, Colorado State University, Fort Collins, 2003.

Kondolf, G. Mathias, Matthew W. Smeltzer, and Steven F. Railsback. "Design and Performance of a Channel Reconstruction Project in a Coastal California Gravel-Bed Stream." *Environmental Management* 28 (2001): 761–776.

Lopez, Barry H. *River Notes: The Dance of Herons.* New York: Avon, 1979.

Wohl, Ellen, Paul L. Angermeier, Brian Bledsoe, G. Mathias Kondolf, Larry MacDonnell, David M. Merritt, Margaret A. Palmer, N. LeRoy Poff, and David Tarboton. "River Restoration." *Water Resources Research* 41 (2005): W10301.

THE DISILLUSIONED ANGLER

Harig, Amy L., and Kurt D. Fausch. "Minimum Habitat Requirements for Establishing Translocated Cutthroat Trout Populations." *Ecological Applications* 12 (2002): 535–551.

Kennedy, Christopher. "History of the Fisheries in the Rocky Mountain National Park Vicinity, with Special Emphasis on the Greenback Cutthroat Trout." Paper presented at the Colorado Riparian Association Sixteenth Annual Conference, Estes Park, Colorado, 2004, pp. 39–44; http://coloradoriparian.org/conferences/con2004/index.php.

Thompson, Douglas M. "Did the Pre-1980 Use of In-Stream Structures Improve Streams? A Reanalysis of Historical Data." *Ecological Applications* 16 (2006): 784–796.

POISONING THE WELL

Adenlof, Katherine A. "Controls on Channel Morphology in a Subalpine Stream." Master's thesis, Colorado State University, Fort Collins, 1992.

Adenlof, Katherine A., and Ellen E. Wohl. "Controls on Bedload Movement in a Subalpine Stream of the Colorado Rocky Mountains, USA." *Arctic and Alpine Research* 26 (1994): 77–85.

Baker, B. W., D. C. S. Mitchell, H. C. Ducharme, T. R. Stanley, and H. R. Peinetti. "Why Aren't There More Beaver in Rocky Mountain National Park?" Paper presented at the Colorado Riparian Association Sixteenth Annual Conference, Estes Park, Colorado, 2004, pp. 85–90; http://coloradoriparian.org/conferences/con2004/index.php.

Baron, Jill S., H. M. Rueth, A. M. Wolfe, K. R. Nydick, E. J. Allstott, J. T. Minear, and B. Moraska. "Ecosystem Responses to Nitrogen Deposition in the Colorado Front Range." *Ecosystems* 3 (2000): 352–368.

Beaty, David W., Gary P. Landis, and Tommy B. Thompson, eds. *Carbonate-Hosted Sulfide Deposits of the Central Colorado Mineral Belt.* Economic Geology Monograph 7. New Haven, CT: Economic Geology Publishing Company, 1990.

Bergen, Gretchen. "Dredging Sediment from the Fraser River." *Denver Post,* February 3, 2008, pp. 1E, 6E.

Blizard, Clifford R. "Hydraulic Variables and Bedload Transport in East St. Louis Creek, Rocky Mountains, Colorado." Master's thesis, Colorado State University, Fort Collins, 1994.

Blizard, Clifford R., and Ellen E. Wohl. "Relationships between Hydraulic Variables and Bedload Transport in a Subalpine Channel, Colorado Rocky Mountains, USA." *Geomorphology* 22 (1998): 359–371.

Cariveau, Alison B., and M. A. Reddy. "Riparian Conservation Project Monitoring and Avian Habitat in Colorado." Paper presented at the Colorado Riparian Association Sixteenth Annual Conference, Estes Park, Colorado, 2004, pp. 23–38; http://coloradoriparian.org/conferences/con2004/index.php.

Colborn, Theo, Diane Dumanoski, and John Myers. *Our Stolen Future: Are We Threatening Our Fertility, Intelligence, and Survival?* New York: Dutton, 1996.

Curran, Janet H. "Hydraulics of Large Woody Debris in Step-Pool Channels, Cascade Range, Washington." Master's thesis, Colorado State University, Fort Collins, 1999.

Curran, Janet H., and Ellen E. Wohl. "Large Woody Debris and Flow Resistance in Step-Pool Channels, Cascade Range, Washington." *Geomorphology* 51 (2003): 141–157.

Dempsey, Stanley, and James E. Fell. *Mining the Summit: Colorado's Ten Mile District, 1860–1960.* Norman: University of Oklahoma Press, 1986.

Dennehy, Kevin F., David W. Litke, Cathy M. Tate, Sharon L. Qi, Peter B. McMahon, Breton W. Bruce, Robert A. Kimbrough, and Janet S. Heiny. *Water Quality in the South Platte River Basin, Colorado, Nebraska, and Wyoming, 1992–1995.* U.S. Geological Survey Circular 1167, 1998.

Donn, J., M. Mendoza, and J. Pritchard. "Feds Look for Effects of Meds-Tainted Fish: Problems with Sex Organs, Other Defects Found." *Rocky Mountain News,* March 11, 2008, p. 20.

Ellis, Anne. *The Life of an Ordinary Woman.* Boston: Houghton Mifflin, 1929.

———. *Plain Anne Ellis.* Boston: Houghton Mifflin, 1931.

Erickson, J. "Pollutants Raining Down on Rockies." *Rocky Mountain News,* August 30, 2004, pp. 5A, 12A–15A.

Fenn, Mark E., Jill S. Baron, Edith B. Allen, Heather M. Rueth, and others. "Ecological Effects of Nitrogen Deposition in the Western United States." *BioScience* 53 (2003): 404–420.

Gordon, Nancy. *Summary of Technical Testimony in the Colorado Water Division 1 Trial.* U.S. Forest Service General Technical Report RM-GTR-270. Washington, DC: U.S. Forest Service, 1995.

Grimm, Michael M. "Paleoflood History and Geomorphology of Bear Creek Basin, Colorado." Master's thesis, Colorado State University, Fort Collins, 1993.

Grimm, Michael M., Ellen E. Wohl, and Robert D. Jarrett. "Coarse-Sediment Distribution as Evidence of an Elevation Limit for Flash Flooding, Bear Creek, Colorado." *Geomorphology* 14 (1995): 199–210.

Harris, R. "Dust Storms Threaten Snow Packs." National Public Radio, Morning Edition, May 30, 2006, www.npr.org/templates/story/story.php?storyId=5415308, accessed on August 27, 2008.

Hilmes, Marsha M. "Changes in Channel Morphology Associated with Placer Mining along the Middle Fork of the South Platte River, Fairplay, Colorado." Master's thesis, Colorado State University, Fort Collins, 1993.

Hilmes, Marsha M., and Ellen E. Wohl. "Changes in Channel Morphology Associated with Placer Mining." *Physical Geography* 16 (1995): 223–242.

Hughes, T. "Intersex Fish Found in RMNP: Report Links Mutation to Air Pollution." *Fort Collins Coloradoan,* February 28, 2008, pp. A1–A2.

Johnson, Therese, and Ryan Monello. "Elk and Vegetation Management in Rocky Mountain National Park: Research and Management Planning." Paper presented at the Colorado Riparian Association Seventeenth Annual Conference, Estes Park, CO, 2004; http://coloradoriparian.org/conferences/con2004/index.php.

Landers, Dixon H., Staci Simonich, Daniel Jaffe, Linda Geiser, Donald H. Campbell, Adam Schwindt, Carl Schreck, Michael Kent, Will Hafner, Howard E. Taylor, and others. *The Fate, Transport, and Ecological Impacts of Airborne Contaminants in Western National Parks*

(USA). EPA/600/R-09/138. Corvallis, OR: U.S. Environmental Protection Agency, Office of Research and Development, NHEERL, Western Ecology Division, 2008.

Lipsher, Steve. "Wildfire Smoke a Culprit in Mercury's Toxic Spread." *Denver Post,* October 19, 2007, p. 5B.

Merritt, David. "The Effects of Mountain Reservoir Operations on the Distributions and Mechanisms of Riparian Plants, Colorado Front Range." PhD diss., Colorado State University, Fort Collins, 1999.

Nadler, C. T., and Stanley A. Schumm. "Metamorphosis of South Platte and Arkansas Rivers, Eastern Colorado." *Physical Geography* 2 (1981): 95–115.

Phelps, Tracy L. "Investigation of the Hydraulic Patterns in a Riffle Using Three-Dimensional Velocity Characteristics." Master's thesis, Colorado State University, Fort Collins, 2003.

Pisani, Donald J. "To Reclaim a Divided West: Water, Law, and Public Policy, 1848–1902." Albuquerque: University of New Mexico Press, 1992.

Pruess, Jonathan W. "Paleoflood Reconstructions within the Animas River Basin Upstream from Durango, Colorado." Master's thesis, Colorado State University, Fort Collins, 1996.

Pruess, Jonathan W., Ellen E. Wohl, and Robert D. Jarrett. "Methodology and Implications of Maximum Paleodischarge Estimates for Mountain Channels, Upper Animas River Basin, Colorado, USA." *Arctic and Alpine Research* 30 (1998): 40–50.

Rathburn, Sara L. "Modeling Pool Sediment Dynamics in a Mountain River." PhD diss., Colorado State University, Fort Collins, 2001.

Rathburn, Sara L., and E. Wohl. "Predicting Fine Sediment Dynamics along a Pool-Riffle Mountain Channel." *Geomorphology* 55 (2003): 111–124.

Reisner, Marc, and S. Bates. *Overtapped Oasis: Reform or Revolution for Western Water.* Washington, DC: Island Press, 1990.

Scanlon, B. "Contaminants Raise Disturbing Questions." *Rocky Mountain News,* March 11, 2008, p. 20.

Simmons, Virginia M. *The Upper Arkansas: A Mountain River Valley.* Boulder, CO: Pruett, 1990.

Smith, J. "Water Crisis Grows Dire amid Dry Skies, Wells: Eastern Plains Farmers Confronted by Uncertain Future as Drought, Cities' Relentless Thirst Threaten Agriculture." *Rocky Mountain News,* July 9, 2007, pp. 6–9.

Stegner, Wallace. *Angle of Repose.* Garden City, NY: Doubleday, 1971.

Stein, Theo, and Miles Moffeit. "Mutant Fish Prompt Concern." *Denver Post,* October 3, 2004, pp. 1A, 18A.

Strom, Sean M. "The Utility of Metal Biomarkers in Assessing the Toxicity of Metals in the American Dipper *(Cinclus mexicanus).*" Master's thesis, Colorado State University, Fort Collins, 2000.

Thompson, Douglas M. "Hydraulics and Sediment Transport Processes in a Pool-Riffle Rocky Mountain Stream." Master's thesis, Colorado State University, Fort Collins, 1994.

———. "Hydraulics and Pool Geometry." PhD diss., Colorado State University, Fort Collins, 1997.

Thompson, Douglas M., Jonathan M. Nelson, and Ellen E. Wohl. "Interactions between Pool Geometry and Hydraulics." *Water Resources Research* 34 (1998): 3673–3681.

Thompson, Douglas M., Ellen E. Wohl, and Robert D. Jarrett. "A Revised Velocity-Reversal and Sediment-Sorting Model for a High-Gradient, Pool-Riffle Stream." *Physical Geography* 17 (1996): 142–156.

———. "Velocity Reversals and Sediment Sorting in Pools and Riffles Controlled by Channel Constrictions." *Geomorphology* 27 (1999): 229–241.

Trayler, Carolyn R. "Spatial and Temporal Variability in Sediment Movement, and the Role of Woody Debris in a Sub-alpine Stream, Colorado." Master's thesis, Colorado State University, Fort Collins, 1997.

Trayler, Carolyn R., and Ellen E. Wohl. "Seasonal Changes in Bed Elevation in a Step-Pool Channel, Rocky Mountains, Colorado, USA." *Arctic, Antarctic, and Alpine Research* 32 (2000): 95–103.

Tyler, Daniel. *The Last Water Hole in the West: The Colorado–Big Thompson Project and the Northern Colorado Water Conservancy District.* Niwot: University Press of Colorado, 1992.

Westbrook, Cheri J., David J. Cooper, and Bruce W. Baker. "Beaver vs. Floods in Controlling Hydrological Processes on Floodplains." Paper presented at the Colorado Riparian Association Sixteenth Annual Conference, Estes Park, Colorado, 2004, pp. 91–94; http://coloradoriparian.org/conferences/con2004/index.php.

Wiedinmyer, Christine, and Hans Friedli. "Mercury Emission Estimates from Fires: An Initial Inventory for the United States." *Environmental Science and Technology* 41 (2007): 8092–8098.

Wilkinson, Charles F. *Crossing the Next Meridian: Land, Water, and the Future of the West.* Washington, DC: Island Press, 1992.

Williams, Garnett P. *The Case of the Shrinking Channels—the North Platte and Platte Rivers in Nebraska.* U.S. Geological Survey Circular 781, 1978.

Wohl, Ellen E. *Virtual Rivers: Lessons from the Mountain Rivers of the Colorado Front Range.* New Haven, CT: Yale University Press, 2001.

———. *Disconnected Rivers: Linking Rivers to Landscapes.* New Haven, CT: Yale University Press, 2004.

Wohl, Ellen E., and Daniel A. Cenderelli. "Sediment Deposition and Transport Patterns Following a Reservoir Sediment Release." *Water Resources Research* 36 (2000): 319–333.

EPILOGUE

Diamond, Jared. *Collapse: How Societies Choose to Fail or Succeed.* New York: Viking, 2005.

Lopez, Barry H. *Arctic Dreams: Imagination and Desire in a Northern Landscape.* New York: Bantam, 1986.

Stegner, Wallace. *The Sound of Mountain Water.* 1946. Reprint, Lincoln: University of Nebraska Press, 1980.

———. *Where the Bluebird Sings to the Lemonade Springs: Living and Writing in the West.* New York: Random House, 1992.

Worster, Donald. *Under Western Skies: Nature and History in the American West.* New York: Oxford University Press, 1992.

Wright, Judith. "Unknown Water." In *Collected Poems.* Sydney, Australia: Angus and Robertson, 1971.

INDEX

Text:	11/15 Granjon
Display:	Granjon
Compositor:	BookMatters, Berkeley
Printer and binder:	Thomson-Shore, Inc.